D1175586

Recombinant DNA

Recombinant DNA
Science, Ethics, and Politics

Edited by

John Richards

Department of Philosophy and Religion
University of Georgia
Athens, Georgia

ACADEMIC PRESS   New York   San Francisco   London   1978
A Subsidiary of Harcourt Brace Jovanovich, Publishers

ACADEMIC PRESS, INC.
111 Fifth Avenue, New York, New York 10003

United Kingdom Edition published by
ACADEMIC PRESS, INC. (LONDON) LTD.
24/28 Oval Road, London NW1   7DX

Library of Congress Cataloging in Publication Data

Main entry under title:

Recombinant DNA : science, ethics, and politics.

Bibliography: p.
Includes index.
1. Recombinant DNA—Congress.   I.   Richards,
John, Date
QH442.R42   574.8'732   78-11355
ISBN 0-12-587480-4

PRINTED IN THE UNITED STATES OF AMERICA

78 79 80 81 82   9 8 7 6 5 4 3 2 1

To Robi

Contents

ETHICS

POLITICS

BEYOND RECOMBINANT DNA

APPENDIX

Contents

List of Contributors

Numbers in parentheses indicate the pages on which the authors' contributions begin.

Tom L. Beauchamp (225), Department of Philosophy and the Kennedy Institute, Center for Bioethics, Georgetown University, Washington, D.C. 20057

Daniel Callahan (135), The Hastings Center, Institute of Society, Ethics, and the Life Sciences, Hastings-on-Hudson, New York 10706

David Clem (241), City Council, Cambridge, Massachusetts

Roy Curtiss III (149), Department of Microbiology, University of Alabama Medical Center, Birmingham, Alabama 35294

Samuel B. Formal (127), Department of Bacterial Diseases, Walter Reed Army Institute of Research, Washington, D.C. 20012

Harold Green (193), National Law Center, Georgetown University, Washington, D.C. 20001

Susan G. Hadden (207), Southern Center for Studies in Public Policy, Clark College, Atlanta, Georgia 30314

Hans Jonas (253), The New School for Social Research, New York, New York 10001

Sidney R. Kushner (35), Program in Genetics, Department of Biochemistry, University of Georgia, Athens, Georgia 30602

Richard Novick (71), Department of Plasmid Biology, The Public Health Research Institute of the City of New York, Inc., New York, New York 10001

John Richards (1, 157), Department of Philosophy and Religion, University of Georgia, Athens, Georgia 30602

Michael Ruse (103), Department of Philosophy, The University of Guelph, Ontario, Canada

Robert L. Sinsheimer (17), Division of Biology, California Institute of Technology, Pasadena, California 91125

Raymond C. Valentine (59), Plant Growth Laboratory, University of California, Davis, California 95616

Mary B. Williams (177), University of Delaware, Dover, Delaware 19901

Burke K. Zimmerman (273), Subcommittee on Health and the Environment, Committee on Interstate and Foreign Commerce, House of Representatives, Washington, D.C. 20515

Preface

The papers in this volume come from two sources. The majority have evolved from papers presented at a conference, *Ethical and Methodological Dimensions of Scientific Research: Recombinant DNA, A Case Study,* held at the University of Georgia, April 15–16, 1977. The conference was cosponsored by the Department of Philosophy and Religion of the University of Georgia and the Atlanta Area American Association for the Advancement of Science. The program was financially assisted by the National Endowment for the Humanities through the Committee for the Humanities in Georgia and the American Association for the Advancement of Science.

In addition to the conference papers, three papers have been solicited to round out the volume. Sidney R. Kushner provides an introduction to the methods and uses of recombinant DNA technology. And Hans Jonas and Burke Zimmerman suggest ways in which the debate can and should be extended to science and scientific research generally. I have also reprinted several reports from the early conference and letters that marked the beginning of the controversy in an appendix. An extensive annotated bibliography is provided as a basis for further consideration of the issues raised in this volume.

Many people assisted in the various stages of this project. I would like to express my thanks to Professor Bruce Carlton of the Department of Biology, University of Georgia, Mr. J. Preston Prather of the Committee for the Humanities in Georgia, and Ms. Helen Mills of the Georgia Center for Continuing Education for their assistance in the planning of the conference. I owe a special debt of gratitude to the late Professor William T. Blackstone for his support and guidance in this project. I would also like to thank Professors Bernard Dauenhauer, Nicholas T. Fotion, and Susan G. Hadden for their advice on the manuscripts, Ms. Joan Hoffman for typing the manuscripts, and Ms. Lucile Epperson for her secretarial assistance.

Introduction

It is clear that science is often in conflict with society
or with powerful groups or vested interests in society.
Sometimes science seems to threaten social stability, at
others to run counter to the dominant aims of society. The
problem is how to reconcile the autonomy of science with the
needs of society as a whole. It is not easy, but it must
be done if we are to enjoy the benefits which science alone
can bring to society.

> -- Julian Huxley, *Soviet Genetics and*
> *World Science; Lysenko and the*
> *Meaning of Heredity,* ix.

The controversy over the regulation of recombinant DNA research
brings cherished societal values into conflict. How do we balance
society's responsibility to protect its citizens with the right of
the scientist to free speech and freedom of investigation? Is the
value of knowledge absolute, or must it be mitigated by considerations of the "public good" and the quality of life? Should scientists assume full responsibility for their activities, or will
society preempt this right and impose regulations? To what extent
can we estimate and predict results in areas of research which
have never before been developed? Where do we locate the burden
of proof in these cases? And what are the implications of stifling human curiosity and creativity?

The comprehensive quality of these issues transcends traditional disciplinary boundaries. We are forced to move beyond the
training and experience of our respective disciplines, and to

partially shed our protective cloaks of expertise. Because of the
overwhelming penetration of these values into every aspect of our
lives, no single discipline is adequate to the task.

In this volume the recombinant DNA controversy provides a
vehicle through which the authors bring their own diverse expertise to bear on these broader issues. In this respect, the volume
itself is a study of the potential role of the expert in complex
decision-making processes which inherently extend beyond any field
of expertise.

By tying these broad issues to a specific case, we breach the
barrier between theory and practice. These are not idle considerations of hypothetical situations. This is a real controversy
which has not yet been resolved. This is not an abstract consideration of the decision-making process, rather it is incumbent upon
us to resolve issues and to arrive at a decision. If this decision
is to hold together, it must ultimately rely on a complex fabric
of expertise--a careful interweaving of the judgment of microbiologists, epidemiologists, safety officers, ethicists, lawyers,
philosophers of science, politicians, and political scientists.
To this end, in this volume, experts from among these diverse
fields have focused on the regulation of recombinant DNA research
as a means to clarify and grapple with these issues which arise
naturally in the interface of science and society.

THE RECOMBINANT DNA DEBATE

Recombinant DNA (rDNA) research involves the addition of one
or more foreign genes to the gene structure of a living organism,
usually the bacterium *Escherichia coli* (*E. coli*). While this
addition is generally quite minimal in terms of the total genetic
material of the host organism, the net effect of the recombination
may be the creation of a radically altered, or even a new, form of
life. Recombinant DNA technology offers hope of a cure for cancer

or the key to developing an unlimited food and energy supply.
This comes at a price: the threat of new and uncontrollable
disease organisms, dangerous to man or to some other presently
existing organism. Because of the unknown and unpredictable
nature of this research, the National Institutes of Health (NIH)
has adopted the *Guidelines for Research Involving Recombinant DNA
Molecules*. These are guidelines for the conduct of experiments
supported by NIH funds. These Guidelines have now been extended
to cover any research supported by federal funds. They establish
a basis for evaluating the comparative danger of the experiments,
and, accordingly, require that different levels of containment be
provided, appropriate to the estimated level of danger. This
containment is of two types. *Biological containment* relies on
using weakened strains of the host organisms so that if they were
to escape the laboratory they would have difficulty surviving.
There are three levels of biological containment: EK1 - EK3.
Physical containment involves the establishment of four levels of
security, P1-P4, which includes laboratory procedures and, at the
P3 and P4 levels, specialized construction requirements. In
addition, there are experiments which the NIH deems too dangerous
to be conducted under any conditions.

The central aspect of the debate concerns our ability to
evaluate benefits and risks when there are essential unknown
factors. The *potential benefits* are great, but none have been
realized yet, and the *potential harm* may also be great, but again
none has been realized. In many ways we are reduced to predicting
the future, and that is rarely productive or accurate. It may be
years before the benefits pass beyond basic research to society,
and if there is harm, it may be years or decades before it is
discovered. Even then it may be impossible to trace it to the
recombination experiments.

The major problem with evaluating the research can be best
understood if we look at the beginnings of the debate. *E. coli*
has been studied and utilized in laboratories for decades. It is

probably the best known organism in the world. It normally
resides in the intestines of humans and other mammals. Because it
is so well known to biologists, it is clearly the host of choice
for recombination experiments. Changes caused by the insertion
of new DNA into *E. coli* would be easier to detect than in a less
well known host organism. In 1971, at Stanford, an experiment was
planned to insert DNA from simian virus 40 (SV 40) into *E. coli*.
SV 40 is extremely simple, having only a few genes. It was isolated initially in monkeys where it resides benignly. When it is
injected into mice or hamsters, it is known to cause cancer. The
goal of the experiment was to try to isolate one of the genes in
SV 40 which might help to identify it as the cause of virus-
induced cancer. Thus from the very start, the promise in the
recombination experiments has been great. However, in the structure of this same experiment lay the potential harm. What would
happen if somehow the ability to cause cancer were transferred to
E. coli, and what would happen if the altered *E. coli* escaped the
laboratory and survived? There was no way to predict. The
researchers cancelled the experiment, and over the next several
years a series of conferences were held resulting in the NIH
Guidelines. In the Appendix we have included selected letters and
reports from these conferences which provide a more comprehensive
picture of the early stages of the debate.

In spite of the care taken in their formation, the NIH Guidelines remain controversial. Some researchers claim they are too
stringent, and unnecessarily retard progress. Others argue that
they are too lenient, that the lower physical levels of containment (P1 and P2) are not containment at all, just good laboratory
technique, and that the higher level containment facilities should
be limited, centralized, and closely supervised. Still others
argue that the NIH Guidelines need to be supported by federal
legislation in order to pertain to research not receiving federal
funds.

On a more comprehensive level, the decision process which
resulted in the NIH Guidelines has been challenged. Is the regulation of scientific research properly within the jurisdiction of
NIH? Was there sufficient public input in the formation of the
Guidelines, or was NIH forced to rely too heavily on those
involved in the research? Can, and should, we replace the *ad hoc*
actions which produced first, the moratorium and later the Guidelines, with a more permanent structure such as a science court?

As the scientific community, Congress, and the public seek
equitable solutions to these questions, there is a discernible
need to defuse some of the more emotional aspects of the debate.
We hope to enhance this process by considering the rDNA controversy in its broader context, as a part of the historic, yet
evolving, relationship between science and society.

THE GALILEAN IMPERATIVE

Professor Robert L. Sinsheimer, Chancellor of the University
of California at Santa Cruz, and formerly Chairman of the Division
of Biology at California Institute of Technology, is one of the
leading critics of continuing research under the NIH Guidelines.
He has argued that, by introducing interspecific genetic recom-
binations, we are intervening in the evolutionary process in a
novel way and are then totally unable to predict the consequences
of this intervention. This "evolutionary perspective" demands
more stringent guidelines for rDNA research.

In his opening paper, Sinsheimer argues that contemporary
science has been dominated by the legacy of Galileo's battle with
the Church, the Galilean Imperative: ". . . to explore every domain,
unravel every mystery, penetrate every unknown, explain every process. Consider not the cost, abide no interference, in the holy
pursuit of truth." Sinsheimer challenges the efficacy of this
Imperative, and proposes, instead a longer perspective which

acknowledges the rights of future generations. From this perspective we gain, "a sense of balance which reminds us that we need
ethics and organization and faith as well as knowledge."

This perspective removes the immediacy of discovery. It
counters the tendency of our scientific know-how to exceed our
understanding of what to do with it. Regarding rDNA, the perspective buys time to consider the wisdom of creating novel living
organisms. Is it in fact wise or proper for us to reshape the
world of life to our will? Perhaps it is proper, and perhaps we
may be willing to accept this responsibility, and perhaps we can
proceed with the research. But, if this is our course, Sinsheimer
argues, then we must proceed only with extreme caution, and with
the constant awareness that containment can never be totally
effective.

THE DEBATE

In this part of the volume we turn to the actual debate.
Recombinant DNA technology resulted from advances in several
divergent fields of research spanning modern molecular biology.
Correspondingly, the technology has far-reaching methodological
consequences. In the first paper, Sidney Kushner of the Program
in Genetics, Department of Biochemistry, University of Georgia,
provides a brief historical survey of modern molecular biology.
He traces the development of the seemingly unrelated lines of
research which culminated in rDNA technology. This survey provides a basis for explaining a hypothetical rDNA experiment and
for explicating the potential benefits of this technology for basic
research.

Kushner concludes that potential applications of the research
are far off--as are the much discussed potential risks. Nevertheless, it is only through the pursuit of knowledge that we can
attain the former and avoid the latter.

In Defense of Continuing Research

There are, roughly, four areas in which the benefits of rDNA can be anticipated. First, and most significant, application of the new technology will allow researchers to isolate single genes or clusters of a few related genes from a complex organism. These can be replicated so that large quantities of the gene would be available for experimental purposes, allowing researchers to probe the molecular structure of these individual genes. Prior to this technology, in practically every case, these genes were not available for the type of detailed analysis required. Second, the technology might lead to a greater understanding of genetic defects, by isolating and analyzing particular genes. For example, hemoglobin genes are currently being isolated and analyzed in great detail. This application has great promise as there are over 1500 different diseases associated with genetic defects. Third, by inserting natural or synthetic genes into microorganisms they could be reproduced in a commercially feasible manner. The bacterial cell is, in effect, used as a factory for cloning the genes. For example, if we obtain human insulin in this manner, then this might lead to better management of diabetes than is now possible with non-human insulin. Fourth, this technology may allow researchers to manipulate natural genes in mammalian cells. Genes may be isolated, and then placed back in their natural environment in order to better isolate their specific function.

Raymond Valentine of the Plant Growth Lab, University of California at Davis, discusses the prospects of utilizing the technology to introduce the nitrogen fixing property into corn, wheat, and other cereal grains that now require massive quantities of chemical-nitrogen fertilizer. Every acre of corn needs about 100 pounds of chemical fertilizer which is derivable in many cases from oil. Valentine proposes that plant scientists might relieve, to some small degree, our massive requirement for chemical-nitrogen fertilizer which uses up this very valuable non-renewable resource.

The Dangers of Unrestricted Research

Richard Novick, Chief, Department of Plasmid Biology of the
Public Health Institute of the City of New York, presents the
arguments against unrestricted research. These center around the
release of a novel recombinant organism into the biosphere. An
accidental dissemination of the novel recombinants, would clearly
have unpredictable consequences, resulting, perhaps, in an entirely
unexpected modification of the biology of the host organism. They
might, for example, modify the organism to increase resistance to
antibiotics. On the other hand, if the experiments are successful,
then some of the proposed uses (for example, contolling oil spills
with "oil eating" bacteria) would require an intentional dissemination of novel recombinants. But organisms in new habitats often
cause trouble in unexpected ways, with unpredictable side effects.
These disseminations might be dangerous, for no reason other than
their novelty.

The novelty of the recombinants is somewhat assured by the
nature of their production. Recombination experiments can cross
barriers that are rarely, if ever, crossed in nature. Thus, in
some experiments, there is a breaching of the barrier between
lower organisms, prokaryotes (those which lack a differentiated
nucleus), and higher organisms, eukaryotes (those which contain
"normal"-looking nuclei). Novick argues that, while in some
special circumstances this barrier has been crossed, its crossing
is neither frequent nor general, and that artificial genetic recom-
binations which do cross this barrier entail a very real biohazard.

Michael Ruse of the Department of Philosophy, University of
Guelph, responds that, under the careful containment conditions of

the NIH Guidelines, a case for the permissibility of rDNA research
can be made. The Guidelines provide sufficient safety standards
given, on the one hand, enormous potential benefits and, on the
other hand, epidemiological considerations which tend to minimize
the risk. Ruse maintains a firm dichotomy between pure science
and technology. He argues, against both Hans Jonas (see the final
section) and Sinsheimer, that rDNA research is genuine pure
science. As such it "is a good in its own right because it is a
prime instance of free human inquiry."

In the final paper of this section Samuel Formal, Chief,
Department of Bacterial Diseases, Walter Reed Army Hospital,
examines the potential for pathogenicity and spread of *E. coli*
K-12, the laboratory strain of *E. coli* which is the usual host for
most rDNA experiments. Results at this time are still tentative,
but Formal reports that they have generally been negative. That
is, deliberate attempts to make the strain pathogenic have not
proven successful, and K-12 has shown an inability to survive for
long periods in the human bowel. Nevertheless, Formal argues,
there is still concern that plasmids containing foreign DNA might
transfer to hardier bacteria and thus have an increased chance of
surviving.

ETHICAL PREREQUISITES FOR EXAMINING BIOLOGICAL RESEARCH

Daniel Callahan, Director of the Institute of Society, Ethics,
and the Life Sciences, Hastings-on-Hudson, argues that the moral
problems of rDNA research cannot be resolved by a cost-benefit
analysis, or by the language of 'trade-offs.' Further, that a
methodology based on such traditional values as freedom, justice,
security and survival, and the pursuit of happiness is insufficient because it would ultimately rely on just such a cost-benefit
analysis.

In dealing with an issue where there is no data, Callahan argues, it is necessary to establish broad moral policies about technology in general, and about the hazards which should be accepted for the sake of technological progress. These moral policies would be established by considerations of general questions regarding (1) the extent to which one is willing to accept man's intervention into nature; (2) whether one prefers risk-taking--gambling to improve life--or caution; and (3) whether one prefers to do good or to avoid harm.

Callahan concludes that the burden of proof lay with those who urge a policy to move forward--to attempt to do good, to gamble-- rather than with those who urge caution. This is manifested in Callahan's contention that any decision to move ahead must gain the assent of the public.

Roy Curtiss of the Department of Microbiology, University of Alabama Medical Center, responded that, in addition to the four values enunciated in the beginning of Callahan's paper, we need to consider the value of knowledge.

By way of reacting to the moral policy options, Curtiss relayed his own involvement with rDNA research. Since 1974, Curtiss had been one of the most outspoken critics of rDNA research. At that time he stopped work to construct rDNA in his own laboratory. Shortly thereafter, Curtiss converted his entire staff over to the design, construction and testing of safer strains of *E. coli* K-12. The result, this past year, was a new strain χ 1776. How- ever, shortly before the Georgia Conference, Curtiss became con- vinced that K-12 itself would be a sufficient biological barrier. This marked a significant shift in thinking within the biological community. Curtiss argues that this detailed evaluation is, in itself, the path of caution and avoidance of harm. Moreover, the use of *E. coli* K-12 host-vectors for permissable experiments poses no threat whatsoever to humans or other organisms, with the excep- tion that a careless lab worker may sometimes suffer harm. He concludes that as long as harm to others is minimized or precluded,

as is the case with the NIH Guidelines, the use of rDNA technology
as a research tool "does not constitute a moral issue of any consequence." He notes though, that these Guidelines need to be
supported by federal legislation in order to pertain to all
parties, public or private.

John Richards of the Department of Philosophy, University of
Georgia, argues that Callahan's broad moral policies are too
general to be of use in the resolution of particular controversies.
Controversy arises from a breakdown in broad moral policies and
each case must then be evaluated on its own merits. There already
is a broad moral policy for science generally, which has not proved
effective in resolving the rDNA debate.

Richards argues that there are four characteristics of rDNA
research which must be taken into account in the determination of
policy: Ignorance, Irreversibility, Global Implications, and
Activity. These features of the research suggest that the pace of
research be slowed, and significant funds be diverted to evaluation of the sufficiency of safety standards and the potential for
hazards.

In the final paper of this section, Mary Williams of the
Freshman Honors Program of the University of Delaware, argues that
the fundamental disagreement in the controversy about rDNA is not
about the probabilities of various undesirable side effects, but
rather about the relative importance of two basic ethical principles. Williams maintains that the opponents of rDNA research are
influenced by an individual worth theory of moral obligation,
while the proponents are influenced by a social benefit theory of
moral obligation. The rDNA controversy is thus best understood as
a conflict between the absolute worth and dignity of each individual human being on the one hand, and the overall benefit for
humankind on the other.

REGULATION AND PUBLIC POLICY

Harold Green, of the National Law Center at Georgetown University, opens this section by detailing a lawyer's perspective of the
history of the rDNA controversy to this point. Green notes that,
while there is a strong analogy between rDNA and nuclear power,
the discussion of the rDNA problem has been significantly more
open and candid than that of nuclear power.

Green, turning to the issue of regulation, argues that regulations regarding rDNA were to be expected and that, in terms of
regulation, there is nothing novel or unique about the research.
He rejects appeals to First Amendment rights of free speech, since
interpretations of the First Amendment have drawn a sharp line of
demarcation between pure speech on the one hand and action on the
other. Speech is generally constitutionally protected under the
First Amendment, but action is not.

Susan G. Hadden, a political scientist with the Southern Center
for Studies in Public Policy argues that there is a tendency for
regulatory agencies to be captured by their clientele, that is, to
allocate power to experts to regulate themselves. This can be
averted if we institutionalize a means for evaluating scientific
advances prior to their becoming crises. 'Clientele capture' can
also be minimized by the enactment of detailed and specific legislation as opposed to broad mandates. The NIH Guidelines fall in
the latter category and need to be modified accordingly.

A three tier regulatory body--at the national, state or county,
and institutional levels--can provide a means of ensuring that
research follows a course responsive to the public and that
scientific information is used in policy-making.

Tom Beauchamp of The Kennedy Institute, Center for Bioethics,
and the Department of Philosophy, Georgetown University, considers
whether the freedom of scientists should be curtailed in order
that nonscientists be protected against risk. He argues that the
consent of the public to possibly dangerous experiments is

essential, and that the right to inquiry is forfeited once a clear
and present danger is shown to result. However, such real dangers
have not been demonstrated in the case of rDNA research. Thus,
Beauchamp concludes, government regulation of rDNA research is not
warranted at this time. Scientific research is protected as freedom of inquiry, and the burden of proof lay with those who would
place special liberty-limiting restrictions on the research.

In the final paper of this section, David Clem, a City
Councillor at Cambridge Massachusetts, outlines the events leading
to the creation of a regulatory city ordinance which established
controls for rDNA research within the city limits. These were
prompted by plans to construct a P-3 laboratory at Harvard. The
lesson gained from this series of events, Clem argues, is straightforward: The lay public has a right to be involved in decisions
affecting their general health and safety and, given adequate
information and meaningful power, the lay public can address complex issues and resolve them equitably.

BEYOND RECOMBINANT DNA

In the concluding section we return to the questions raised in
the beginning by Sinsheimer: Are we justified in restricting the
acquisition of knowledge? Can we raise a legitimate challenge to
the Galilean Imperative?

Hans Jonas of the New School for Social Research considers
rDNA research as indicative of changes occurring within the structure of modern science itself. He argues that modern science
fuses theory with practice, not merely in the expectation of
eventual applications, but in the methodological decision to wrest
knowledge from nature by actively operating on it. Observation
becomes experimentation, which intrinsically intervenes and
manipulates. The freedoms of thought and speech, by which

observation is protected, do not extend to action. Hence, experimentation remains subject to legal and moral restraints.

The very means of acquiring knowledge, Jonas argues, raises
moral questions. In the case of rDNA research, the experiment
itself may become definitive reality. While we are unable to
avoid this inherent fusion of theory with practice, it is possible
to release some of the pressure for progress. This pressure is
only enhanced by pursuing application-oriented research where there
is added incentive for haste and incaution. We can, therefore,
retard the tempo of research by focusing experiments, instead, on
the benefits for basic science and the acquisition of knowledge.

In the final paper of the volume, Burke Zimmerman, formerly
with the Environmental Defense Fund, and presently a staff member
of the House Subcommittee on Health and the Environment, argues
that the debate, to this point, has been highly polarized,
generally inarticulate and unproductive, with most of the participants uninformed in at least one important aspect. Consequently,
for all the flurry of activity, few have been able to free themselves from emotional bias and few of the controversies have been
resolved.

Nevertheless, Zimmerman argues, the one important message which
emerges from this debate is that we must maintain a fundamental
distinction between science--the acquisition of knowledge--and
technology--the application and use of that knowledge. The pursuit of knowledge is a sterile endeavor, and historically, the
suppression of this pursuit has always failed. Zimmerman concludes
that we must heed the caveat that society must not, by confusing
science with technology, or through blundering or ignorance,
needlessly repress the acquisition of knowledge because it fears
the possible application of that knowledge.

SCIENCE, ETHICS, AND POLITICS

The rDNA debate has been carried out almost exclusively in the
scientific sector of the community. The nature of the subject
matter has been such as to exclude the public and the academic
humanist. It is the purpose of this volume to broaden the scope
of the dialogue. The structure of the volume is directed towards
fostering the maximum amount of communication between biologists,
philosophers of science, ethicists, lawyers, political scientists,
politicians, and the general public. By engaging the active
scientist in the field, the theoretician profits by focusing on a
real issue. They, in turn, contribute expertise and a perspective
obtained from their own discipline. This interplay yields a
degree of generality which may permit the development of a wisdom
which can reconcile the needs and responsibilities of society with
the autonomy of science. As is argued in the opening quotation
from Julian Huxley, this is not easy, but it must be done if
science is to really benefit society.

July 1978   John Richards
Athens, Georgia

The Galilean Imperative

Robert L. Sinsheimer*

Division of Biology
California Institute of Technology
Pasadena, California

In Bertolt Brecht's famous play "Galileo," in scene 7, the monk who becomes Galileo's disciple is gripped by doubt. Wracked with anxiety and concern he questions Galileo:

My parents are peasants in the Campagna who know about the cultivation of the olive tree and not much about anything else. . . . I can see the veins stand out on their toil-worn hands and the little spoons in their hands. They scrape a living and underlying their poverty is a sort of order. . . . My father did not get his poor bent back all at once, but little by little, year by year in the olive orchard. . . . They draw the strength they need to sweat with their loaded baskets up the stony paths, to bear children, even to eat, from the sight of the trees greening each year anew . . . and from the little church and the bible texts they hear there on Sunday. They have been told that God relies upon them and that the pageant of the world has been written around them,

*Present address: Office of the Chancellor, University of California at Santa Cruz, Santa Cruz, California.

that they may be tested in the important or unimportant parts
handed out to them. How could they take it were I to tell
them they are on a lump of stone ceaselessly spinning in
empty space, circling around a second-rate star? . . .

I see them slowly put their spoons down on the table.
They would feel cheated . . . they would say "nobody has
planned a part for us beyond this wretched one on a worthless
star; there is no meaning in our misery."[1]

To this, Galileo replies:

Virtues are not exclusive to misery . . . Today the virtues
of exhaustion are caused by the exhausted land. For that my
new water pumps could work more wonders than their ridiculous
superhuman efforts . . .

Shall I lie to your people? . . .

How can new machinery be evolved to domesticate the river
water if we physicists are forbidden to study, discuss and
pool our findings about the greatest machinery of all, the
machinery of the heavenly bodies . . .

You talk of the Campagna peasants . . . I can see their
divine patience, but where is their divine fury?[1]

Of course, these are Brecht's words, not Galileo's. But in
this scene, Brecht exposes the basic quandaries regarding the
human purpose of science, the primal dilemma that the fruit of
knowledge may indeed be bitter, the ancient conflict between truth
and compassion.

But also I would suggest this scene implicitly perpetuates two
deep fallacies, long sanctified by association with Galileo's

[1] From *The Modern Repertoire*, Series Two, edited by Eric Bentley.
Copyright 1952 by Eric Bentley. Reprinted with permission of Indiana
University Press.

martyrdom; first, that science or knowledge is, and must be, of
one piece, indivisible and unbounded--here, that Galileo's discovery of the moons of Jupiter was a necessary accompaniment to
the development of better methods of irrigation. I think that
claim might at least be challenged.

And secondly, that Galileo claimed--and therefore all scientists should claim--the right to unimpeded pursuit of knowledge.
I think more accurately Galileo claimed the right to challenge
accepted human dogma--to burst the mental shackles of a frozen
order--to counter faith in proclaimed truth with empirical fact.
And for this he *was* martyred. But this claim is subtly different
from a claim of absolute freedom to seek new knowledge in every
domain--in unknown worlds concerning which no presumed knowledge
or dogma even exists. While we cannot abide untruth, we might,
with sufficient cause, prefer simple ignorance--at least for a
time.

Such a suggestion is, of course, pure heresy to a scientist,
if not to most of Western thought. It is only very recently, in
this vexed and unraveling time, in a culture disjointed and reeling
from the accelerating impact of science-based change, that such a
voice can even be recognized--that we begin to see that there may
be, if not forbidden knowledge, at least inopportune knowledge--
that just as in the development of an organism the genetic information must be revealed in a progressive order or disaster will
ensue, so in a civilization the accretion of knowledge must be
measured and accompanied by the creation of a cultural context to
control its expression--or disaster may ensue.

We each become the prisoners of our perspective, entrapped in
our expertise.

Scientists believe that the basic human problem is ignorance--
and therefore that the solution to the human dilemma is knowledge--
knowledge of ourselves, knowledge of the universe in which we find
ourselves, knowledge of how we came to be what we are.

Others believe that the basic human problem lies in the use that we make of our knowledge--the decisions we make with it, the ends to which we apply it--and therefore that the solution lies in ethics.

Others believe that, having decided what to do with our knowledge, the basic human problem is how to organize ourselves to achieve our ends in the face of the caprice of Nature and the vagaries of men--and thus that the solution lies in political and social organization.

Still others believe that the solution to the human dilemma is not to be found within ourselves, and thus that the solution is a faith in a higher order.

Unfortunately, it is all too human for those who adopt one perspective or another gradually to forget and even scorn the others. And thus, just as a business may come to believe in profit at any cost, or the military in victory at any cost, so a government may come to believe in sovereignty at any cost, a religion may come to believe in conversion at any cost, and science may come to believe in new knowledge at any cost.

It is unusual, but not incredible today, to suggest that all these perspectives may have some merit--and that, in that perspective, we are becoming dangerously out of balance--that our scientific knowledge, ever growing and mounting in power, is fast exceeding our understanding of what to do with it and the capacity of our social organizations to contain its consequences.

There are two general reasons for this: one is simply the magnitude of current scientific activity. The growth of science has been exponential for many decades. It is estimated that some 90% of all the scientists who ever lived are now alive. The literature of science doubles every ten years. We have--culturally, not biologically--cloned Galileo, a million-fold. And consideration of that reality may give us a premonition of the consequences of biological cloning should that become possible.

Even more important, science has in this century penetrated to
the core of matter and life and in so doing has given us tools of
unprecedented power. Splendid and cumulative discoveries in
physics and chemistry have provided us with a definitive understanding of the nature of matter. From that understanding has
come the technology to reshape the inanimate world to human purpose. And now the description of life in molecular terms provides
the beginnings of a technology to reshape the living world to
human purpose, to reconstruct our fellow life forms--and even our-
selves--into projections of the human will.

It is the release of nuclear energy that has now forced the
first basic reconsideration of the relation of science to human
welfare. The mushroom cloud tells us indelibly that the net
result of scientific progress may not always be benign.

It is instructive to look back to see whether this consequence
was anticipated. In fact, it was.

In a book of essays entitled *Science and Life* published in
1920 by Frederick Soddy, who had been a collaborator of Ernest
Rutherford, we find:

Let us suppose that it became possible to extract the energy,
which now oozes out, so to speak, from radioactive materials
over a period of thousands of millions of years, in as short
a time as we pleased. From a pound weight of such substance
one would get about as much energy as would be obtained by
burning 150 tons of coal. How splendid! Or, a pound weight
could be made to do the work of 150 tons of dynamite. Ah!
there's the rub. . . . It is a discovery that conceivably
might be made tomorrow, in time for its development and perfection for the use or destruction, let us say, of the next
generation, and which, it is pretty certain, will be made by
science sooner or later. Surely it will not need this last
actual demonstration to convince the world that it is doomed,

if it fools with the achievements of science as it has fooled
too long in the past. (Soddy, 1920, p. 36)

War, unless in the meantime man had found a better use
for the gifts of science, would not be the lingering agony it
is today. Any selected section of the world, or the whole of
it if necessary, could be depopulated with a swiftness and
dispatch that would leave nothing to be desired. (Soddy,
1920, p. 107)

Similarly, in 1921 Walther Nernst, the famous German physicist,
discussing the possibility of triggering the sudden release of the
energy latent in uranium, suggested that "We may compare the existence of mankind somewhat with that of a primitive people who live
on an island composed of gun cotton, but are not in possession of
fire. This colony would vanish in the instant in which Prometheus
handed them a torch" (Nernst, 1921, p. 24).

What was the response to these warnings? Nonsense!
In 1933 Rutherford himself said:

It has sometimes been suggested, from analogy with ordinary
exposions, that the transmutation of one atom might cause the
transformation of a neighboring nucleus, so that the explosion
would spread throughout the material. If that were true we
should long ago have had a gigantic explosion in our laboratories with no one remaining to tell the tale. . . . On the
average we could not expect to obtain energy in this way. It
is a very poor and inefficient way of producing energy and
anyone who looks for a source of power in the transformation
of atoms is talking moonshine.[2]

[2]The British Broadcasting Corporation's National Lecture,
June, 1933, as quoted in Evans (1939), p. 201.

This was thirteen years after Soddy's book--and twelve years
to Hiroshima.

Similarly, in the same period, Robert Millikan, the famous
American physicist then President of Caltech, wrote in an article
entitled "The Alleged Sins of Science":

> Since Mr. Soddy raised the hobgoblin of dangerous quantities
> of available subatomic energies [science] has brought to light
> good evidence that this particular hobgoblin--like most of the
> hobgoblins that crowd in on the mind of ignorance--was a myth.
> . . . The new evidence born of further scientific study is to
> the effect that it is highly improbable that there is any
> appreciable amount of available subatomic energy for man to
> tap. (Millikan, 1930, p. 121)

So much for scientific prophecy.

Later in the same article Millikan, with a rather touching
faith, wrote, one may "sleep in peace with the consciousness that
the Creator has put some foolproof elements into his handiwork and
that man is powerless to do it any titanic physical damage"
(Millikan, 1930, p. 121).

We can no longer have *that* faith. But it is indeed instructive, and also troubling, to recognize that our scientific endeavor
truly does rest upon unspoken, even unrecognized faith--a faith in
the resilience, even in the benevolence, of Nature as we have
probed it, dissected it, rearranged its components in novel con-
figurations, bent its forms, and diverted its forces to human purpose. The faith that our scientific probing and our technological
ventures will not displace some key element of our protective
environment, and thereby collapse our ecological niche. A faith
that Nature does not set booby traps for unwary species.

Our bold scientific thrusts into new territories, uncharted by
experiment and unencompassed by theory, must rely wholly upon such
faith. In the past that faith has been justified and rewarded,

but will it always be so? The faith of one era is not always
appropriate to the next, and an unexamined faith is unworthy of
science. Ought we step more cautiously as we explore the deeper
levels of matter and life?

Even less overtly, we have relied upon a faith in the
resilience of our social institutions--in their capacity to contain the stress of change, and to adapt the knowledge gained by
science--and the power inherent in that knowledge--to the benefit
of man and society more than the detriment. That faith, too, is
increasingly strained by the acceleration of technical change and
the magnitude of the powers deployed.

Physics and chemistry have given us the power to reshape the
physical nature of the planet. We wield forces comparable to, even
greater than, those of natural catastrophes. And now biology is
bringing to us a comparable power over the world of life. The
recombinant DNA issue, while significant and potentially a
grievous hazard in itself, must be seen as a portent of things to
come.

Genes determine the basic structures and biological potentials
of all living forms. Through the reconstruction, the realignment,
the recombination of genes, we can in time reshape the nature of
all life on earth, including, in ultimate paradox, our own.

The immediate issue of recombinant DNA concerns the wisdom of
creating novel living organisms, microorganisms, whose containment
cannot be assured and whose release into our biosphere is therefore inevitable.

The scientific value of this research is unquestionable.
Clearly the advent of synthetic biology--of the capacity to isolate specific genetic elements, to try out new genetic combina-
tions--is a most powerful tool for the understanding of life pro-
cesses--as well as a potential source of very significant applications.

But the properties of these novel organisms are essentially
unpredictable and their release into the biosphere is an

extraordinary, unprecedented intervention into the entire process of biological evolution.

From this perspective the invention of synthetic biology is a novel event in the history of science--much as was the evolution of the first living cell a novel event in the abiotic world. For such organisms are self-reproducing. Once released they will, if they find a suitable niche, propagate. They will, in turn, evolve according to their own destiny, wholly beyond our control. Their impact upon the intricate web of life which arose in measured step-by-step interactions over the eons, and which bore and sustains us, is simply incalculable.

Now, again, to these concerns many answer, nonsense. They argue that the existing web of life is too tightly woven and will stifle these new forms. They argue that our intervention can only be but a small perturbation. They argue that we are now suffici-ently advanced so as to be able to cope with any new hazards--that we are in effect secure in our ecological niche.

Of course, most species that have ever lived have perished.

The yea sayers may, in fact, be correct. We may be more secure than we can have reason to know. But can we continue to rely upon that unspoken faith? Is that the course of wisdom?

Continuing genetic research will surely lead to greater and greater power to reshape the world of life--crops, animals, man. With crops and animals we may, in fact, hope to have more control than with microorganisms--although the long-range results of our intervention may still be unpredictable. But another question arises: is it proper or wise for us to reshape the world of life to our will? Each of our fellow passengers on our planet is, like us, the outcome of three billion years of evolution and each carries the potential of future evolution. Ought we assume the responsibility to extinguish this line and invent another? Per-haps we answered that question when we invented agriculture, but the issue is now more stark and more final.

We are becoming creators, inventors of novel forms that will live on long after their makers and will evolve according to their own fates. Before we displace the first Creator we should reflect whether we are qualified to do as well. We may come to know the sorrow of creators.

This issue merges with the broader issue of our obligation to our descendants. As we rearrange the planet and consume its resources and imperiously reshape its flora and fauna, how do we weigh the interests of future generations? The asymmetry of time inevitably frustrates and truncates our ethical commitment to democratic process. Our descendants have no voice and no vote. They need an advocate and a place in our morality.

Recombinant DNA research is not human genetic engineering--but unfortunate experience with other technologies requires us to view recombinant DNA as the first step of a possibly irreversible progression.

The application of genetic engineering to mankind now or in the near future would, I believe, strain the resilience of our social institutions beyond the breaking point. We have through repeated trial and error empirically evolved social forms more or less adapted to the existing gene pool of humanity. We are, as yet, largely ignorant of the character of that gene pool or the essence of that adaptation. But almost certainly any marked perturbation of the human gene pool could have chaotic social consequences in the present or foreseeable future.

There is also, one suspects, a profound innate wisdom in the diversity of the extant human gene pool, evolved and refined by trial and error in the crucible of natural selection. Myriads of our ancestors suffered bitterly to bequeath us this treasure. Diversity is the source of creative interaction and the seed of future adaptation. We should not lightly replace such an intricate pattern with the designs of human ingenuity.

This is not to say Nature cannot be improved. Nature is callous and the price of genetic defect statistically associated

with genetic diversity, need not be borne. Nor is it obvious that
the present human gene pool, less and less subject to the rigors
of natural selection, will continue to remain well adapted as we
evolve, culturally, farther and farther from the primitive conditions that spawned our genome.

There is, let us admit, a profound temptation toward genetic
engineering of all kinds, not merely because it can be done, but
because it would represent a cosmic turnabout--the dethroning or
civilizing, the education of DNA. As the Earth was once displaced
from the center of the universe, we could in turn displace DNA
from its central role as the determinant of life in all its forms.
DNA has served life well--but it has imposed a blind and ruthless
tyranny, a one-eyed obsession with genetic survival. Survival is,
of course, necessary, but is it sufficient?

In Darwinian evolution use does not affect the design. It is
a one-way process, without feedback. Through the application of
intelligence evolution can be converted into an instructive process, a Lamarckian cycle.

Such a cycle, relying upon intelligence, is more easily broken.
The Darwinian process is more conservative and more stable. Once
launched upon genetic engineering we are forever on our own. Dare
we introduce such an historic change?

It is perhaps poetic that the science of genetics should pose
us this terrible challenge. It is in our genes that we have the
potential for reason, the potential for science, the potential
for greatness. It is also in our genes that we have the potential
for madness, for cruelty, for blind passion and base thirst for
power, for hubris. We are ever in a struggle between these aspects
of humanity and as our powers grow the race quickens. We need
consider whether a given advance may arm one side much more than
the other. Tragically, destruction can be so much easier than construction.

The potential for social hazard implicit in the new understanding of biology is not limited to genetic manipulations. Other

lines of research pose deep dilemmas.   In particular, it seems to
me, that research upon the aging process carries the seeds of
great social upheaval and a net social detriment.

   We have--reluctantly, even grimly--come to terms with our
mortality, our allotted span--indeed to the degree that it is the
fact of death that provides meaning for much of life.   To retard
death--not by years but by decades or even centuries--would surely
demand the most profound reconstruction not merely of our social
institutions but of our basic value structures.   The wisdom and
the poetry of the human past would become irrelevant.   For this,
we are grossly unprepared.

   Nor can concern be limited to biological science.   Are attempts
to detect or communicate with postulated extraterrestrial intelli-
gences desirable at this time?   Could we really manage an effective
capacity for weather modification?   Would knowledge of genetic
pre-determination really benefit the affected individuals?

   Some of my colleagues, not only in biology but in other fields
of science as well, have indicated to me that they too increasingly sense that our curiosity, our exploration of Nature, may
unwittingly lead us into an irretrievable disaster.   But they
argue we have no alternative.   Such a position is, of course, a
self-fulfilling prophecy.

   I see no reason to be committed to such a fatalistic stance.
I would prefer to try to use our reason to foresee potential areas
of danger and use the greatest caution; to try to see science not
as an autonomous and heedless juggernaut, but as a part of the
human endeavor, a part of unquestioned import and worth, but conceivably just now a part somewhat out of balance with other parts
of great worth.

   Some will argue that knowledge simply provides us with more
options and thus that the decision point should not be at the
acquisition of knowledge but at its application.

   I suggest that such a view, however ideal, overlooks the
difficulty inherent in the restriction of application of knowledge,

once available in a free society. Does anyone really believe, for
instance, that knowledge permitting an extension of the human life
span would not be applied once it were available?

One must also recognize that the very acquisition of knowledge
can change both the perceptions and the value of the acquirer.
Could, for instance, deeper knowledge of the realities of human
genetics affect our commitment to democracy?

It may be argued that the cost, however, it may be measured, of
impeding research would be greater to a society than the cost of
impeding application. Perhaps so. This issue could be debated.
But it must be debated in realistic terms with regard for the
nature of real people and real society and with full understanding
that knowledge is indeed power.

Indeed, that inevitable conjunction of knowledge and power is
one root cause of our dilemma. The thirst for knowledge is also
for man a thirst for power. Many see human genetic engineering as
what "they" will do to "us."

Gods are all-knowing and as we know more, we approach God
status. But in our genes we are not gods. The Greek gods suggest
what humans would be like, given God-like powers. It is not an
inspiring scene.

Toward the end of the play, Brecht's Galileo, now aged, says,

I take it that the intent of science is to ease human existence.
If you give way to coercion, science can be crippled and your
new machines may simply suggest new drudgeries. Should you
then in time discover all there is to be discovered, your progress must become a progress away from the bulk of humanity.
The gulf might even grow so wide that the sound of your cheer-
ing at some new achievement would be echoed by a universal howl
of horrow.[1]

But as we know all too well, coercion is not needed. A blend
of ambition, or even patriotism, combined with tunnel vision, is
more than adequate. Lace with an undoubting faith in progress, to
provide satisfaction and immunity from conscience, and we have the

Galilean imperative: to explore every domain, unravel every mystery, penetrate every unknown, explain every process. Consider not the cost, abide no interference, in the holy pursuit of truth.

I suggest again that this is a misinterpretation. The Galilean imperative is a challenge, not a command. It is a challenge to the human intellect always to question, to accept as fact only that which has been demonstrated, and then to retain a shadow of doubt. It is not a command to unlock every secret.

That is another imperative: human curiosity, writ large. In our time we can see that human curiosity, like every other passion, must learn restraint. We can see that scientific experiments, some modes of research, may in themselves be too dangerous. And we can see that some kinds of knowledge may be inopportune and may be disastrous for our present social development.

In this age the only coherent, and therefore compelling, vision of the future seems to be that of an increasingly man-designed world, and now one in which even the living forms about us will be genetically sculptured to human purpose. This is the vision that informed the authors of *Brave New World* and *1984*.

This vision is rooted in the human determination to be the masters of our fate—in the end, to leave naught to chance, to the random workings of natural law, nor even to the varied happenstance of that biological evolution which, lest we forget, bore us. For all of this we will substitute human design in the service of human will.

For those whose experience with the expression of human will has not always been cause for celebration, this vision is chilling. But then, *they* must devise another vision equally coherent and compelling. Another vision which will not repudiate the most significant human achievement of the past four centuries, the scientific endeavor. Another vision which understands that to replace chance by choice, brings inevitably a future burden of decision. That to replace the compensated balances of Nature with the thrust of human design places the survival of humanity,

perhaps of all life, thenceforth in human hands. And that blindly
to extol the human intellect is to succumb to hubris.

Through science we touch the eternal. And in the intoxication
of that contact we forget that science too, in the end, is a human
creation, a reality filtered through the human intellect. We can
never know how complete is our perception, how enduring our con-
structs.

And our technology, however brilliant and entrancing, is but
the application of science to the fulfillment of human will and
ambition.

And the loop closes upon us in yet obscure ways because the
society we construct and inhabit inevitably participates in the
shaping of human qualities and their expression.

We should not then make irreversible commitments based upon
ever-partial insights. We should use our powers to ameliorate and
compensate, to brighten and ennoble, not to reconstruct the world
we inherit and bequeath.

We need a more human-scale vision based on a harmony with
Nature rather than absolute dominion over it. An ethos which
continues to explore but which is realistically alert to the
possibility of natural or social disaster. A longer perspective
which recognizes there are many centuries and many generations
ahead. A sense of balance which reminds us that we need ethics
and organization and faith as well as knowledge.

We have culturally cloned Galileo a million-fold. And thereby we have indeed calculated the machinery of the heavens and
even the machinery of the atoms. The first has led us to guided
missiles, the second to nuclear weapons.

We have indeed designed the pumps and irrigated the fields--
and now we face the new cruelties of over-population and the
exhaustion of the resources of the planet.

Today we seek to calculate the machinery of the cell and the
functional trajectories of the genes. Unless we take heed, to
what end will this knowledge be put and with what consequence?

We have nurtured this Galilean clone well; we award prizes and honors to those most like the original. No doubt this clone has served humanity well. But perhaps there is a time for Galileos. Perhaps we need at this time, another clone. Let us think on it.

REFERENCES

Brecht, Bertolt. 1947.   *Galileo--A Play by Bertolt Brecht.*
    English version by Charles Laughton. New York: Grove Press,
    1966.
Evans, Ivor B. N. 1939.   *Man of Power: The Life Story of Baron*
    Rutherford of Nelson. London: Stanley Paul and Co.
Millikan, Robert A. 1930.   "Alleged Sins of Science." *Scribners*
    Magazine, 87, 119-130.
Nernst, Walther. 1921.   *Das Weltgebaude Im Lichte: Der Neuren*
    Forschung. Berlin: Julius Springer.
Soddy, Frederick. 1920. *Science and Life.* London: John Murray.

Science

The Development and Utilization of Recombinant DNA Technology

Sidney R. Kushner

Program in Genetics
Department of Biochemistry
University of Georgia
Athens, Georgia

The ability to construct recombinant DNA molecules (fragments of DNA not normally associated which have been spliced together) in the test tube has resulted from the efforts of many scientists working in seemingly unrelated fields. The nature in which this new technique was developed less than 25 years after the basic structure of the DNA molecule was unraveled, provides some interesting insights into the dizzying rate of advancement in the field

of molecular biology. It is therefore of some interest to provide a brief historical perspective before a detailed explanation
of the methodology is presented.

I. THE DEVELOPMENT OF MODERN MOLECULAR BIOLOGY

The story begins in 1869 when a Swiss biochemist named
Fredrich Miescher isolated a substance called "nuclein." At the
latter part of the century he showed that sperm nuclei consisted
of approximately 60% nucleic acid and 35% protein-like compounds
(as cited in Taylor, 1965). However, as late as the 1920s the
prevalent view was that biological specificity resided in proteins and not nucleic acids. It was not until 1944 that conclu-
sive evidence was presented that deoxyribonucleic acid (DNA) provided the means by which genetic information was transmitted
(Avery *et al.*, 1944). The ability to transfer genetic characteristics was completely destroyed if the material were specifically
degraded. Hotchkiss (1949) showed that upon further purification
the DNA could more efficiently transfer the traits it contained.
Additional proof that DNA was the genetic material was provided
by Hershey and Chase (1952), who demonstrated that upon infection
of a bacteria by a virus only the DNA entered the cell.

In the meantime a number of individuals were attempting to
characterize the chemical composition of DNA. Chargaff and his
colleagues utilized specific reagents which degraded DNA to aid
in the determination of its basic components (1950). At the same
time other investigators were probing the three dimensional
structure by examining the patterns made by X-rays which had
passed through crystals of DNA (Wilkins *et al.*, 1953); Franklin
and Gosling, 1953). Watson and Crick (1953a) analyzed all the
available data and brilliantly deduced the three dimensional
structure of the DNA molecule. This work and their subsequent
paper describing the genetical implications of the structure

(Watson and Crick, 1953b) were to revolutionize the field of
molecular biology.

The deoxyribonucleic acid molecule was shown to consist of
basic building blocks called nucleotides. Each of these nucleotides contained a phosphate group (PO_4) bound to a deoxyribose
sugar (Figure 1).

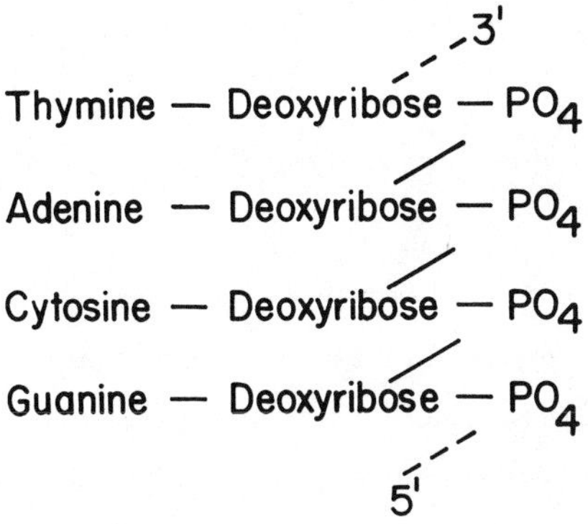

$$-3'$$

Thymine — Deoxyribose — PO_4

Adenine — Deoxyribose — PO_4

Cytosine — Deoxyribose — PO_4

Guanine — Deoxyribose — PO_4

$$5'$$

Figure 1.   The Basic Components of DNA.   Each nucleotide is
            composed of a phosphate group (PO_4), a deoxyribose
            sugar and either a purine (adenine or guanine) or
            pyrimidine (thymine or cytosine) base.

Attached to each deoxyribose sugar was one of four bases [two
pyrimidines (cytosine and thymine) and two purines (adenine and
guanine)] (Figure 1).   The nucleotides were joined together
through linkages attaching the phosphate groups (phosphodiester
bonds) (Figure 1).   Polynucleotide chains (strands) were able to
interact through specific pairing of the purine and pyrimidine
bases (hydrogen bonding).   Thus in naturally occurring double-
stranded DNA molecules adenine (A) was always found associated with
thymine (T), while cytosine (C) always associated with guanine
(G) (Figure 2).   Additionally, the two strands had opposite
polarities (determined by whether the nucleotide at the end of a
chain had a free 3' or 5' group on the deoxyribose sugar, (Figure
2).

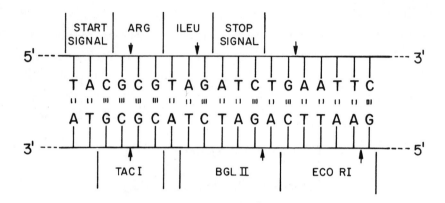

Figure 2.   Structural Features of a DNA Molecule. The two poly-
            nucleotide chains are held together by specific pair-
            ing between the purine and pyrimidine bases (A with T
            and G with C). Groups of three nucleotides provide
            the basis of the genetic code as indicated in the upper
            strand. TAC represents a start signal which tells the
            protein synthesizing machinery of the cell where to
            start reading the code. ATC indicates when to stop.
            Shown on the lower strand are recognition sequences for
            three restriction endonucleases. The arrows indicate
            where the phosphodiester backbone is cleaved by these
            enzymes.

     This specific pairing of nucleotides (complementarity) turned
out to have tremendous significance in terms of the transmission of
genetic information. Watson and Crick (1953b) hypothesized that
nucleotide pairing could provide the means by which DNA molecules
were reproduced (replicated). Messelson and Stahl (1958) were
able to show that inside a cell each of the polynucleotide strands
formed a template for the synthesis of its complement, generating
the two intact DNA molecules necessary when cell division took
place. Kornberg and his colleagues (1956) demonstrated the

template directed synthesis of DNA in a test tube using an extract
of the bacteria *Escherichia coli (E. coli)*.

   The next question to be answered was how the information
encoded on the DNA was converted into the proteins and other
structural components of the cell. Another form of nucleic acid,
called ribonucleic acid, had also been discovered inside cells.
It turned out to have the same basic structure as DNA except that
the deoxyribose sugars were replaced by ribose sugars. In 1961
Jacob and Monod suggested that a short-lived molecule of RNA
(messenger RNA) could provide the intermediate step between the
DNA and the final products. Brenner *et al.* (1961) showed that the
messenger RNA carried the information from the gene to the
protein synthetic machinery of the cell.

   Even with the discovery of messenger RNA, it was still not
clear how only four different nucleotides could provide sufficient information to initiate and control the synthesis of proteins
containing at least 20 different amino acids. Crick first proposed in 1958 that the sequence arrangement of nucleotides provided the code. Nirenberg and Matthaei (1961) provided the
experimental approach to solve this problem by developing an *in
vitro* system for carrying out protein synthesis. Following this
work several laboratories proceeded to demonstrate that groups
of three nucleotides (triplet codon) provided the information
necessary to code for the various amino acids (Figure 2). Additional triplet combinations controlled the initiation and cessation of protein synthesis (Figure 2). Hence although it took
fifty-six years from the first description of "nuclein" to
unravel its three dimensional structure, the genetic code was
cracked in less than 10 years.

II. THE BASIS OF RECOMBINANT DNA TECHNOLOGY

At the same time that these breakthroughs were being made, the
groundwork for recombinant DNA technology was already being laid.
To better understand how all the necessary tools were developed
almost simultaneously, it is appropriate to consider the prerequi-
sites for successfully carrying out such experiments.

A. *Isolation and Purification of DNA*

Before any recombinant DNA molecules could be constructed, it
had to be possible to obtain purified DNA that was sufficiently
large to encode for entire genes. Since DNA naturally exists as
a rod type structure it is particularly sensitive to breakage
during purification. By 1961 procedures for the isolation and
purification of high molecular weight DNA from bacterial and other
sources became available (Marmur, 1961).

B. *Availability of Appropriate Recipients (Hosts)*

As far as recipients were concerned the requirements were
simple. Find an organism that was easy to grow and maintain, and
genetically well-defined. Since scientists had already devoted a
great deal of effort into unraveling the genetics and molecular
biology of the enteric bacteria, *E. coli,* it was a logical choice.
By 1964 over one hundred genes had already been identified in
this organism (Taylor and Thoman, 1964). Since then many more
genes have been characterized (Bachmann *et al.*, 1976). A variety
of means were available by which these genetic determinants could
be transferred from one strain to another.

C. *Production of DNA Fragments which Can Reassociate with Other*
    DNA Fragments Irrespective of Origin

The ability to produce fragments which could reassociate (reanneal) with each other, however, presented a more difficult problem. Reassociation requires the presence of complementary
sequences at the ends of the molecules. Thus the molecules shown

in A and C (see Table 1) could reanneal to form those shown in B
and D, respectively. Those shown in E and F would not reanneal.

Table 1

A	5'	AAAAATACGTCGATAG	3'	5'	TTTTTCGGGCTA		3'
	3'	ATGCAGCTATCAAAAA	5'	3'	GCCCGATTTTTT		5'
B		5'	AAAAATACGTCGATAGTTTTTCGGGCTA		3'		
		3'	ATGCAGCTATCAAAAAGCCCGATTTTTT		5'		
C	5'	AGCTTACGTCGATAG	3'	5'	AGCTCGGGCTA		5'
	3'	ATGCAGCTATCTCGA	5'	3'	GCCCGATTCGA		3'
D		5'	AGCTTACGTCGATAGAGCTCGGGCTA		3'		
		3'	ATGCAGCTATCTCGAGCCCGATTCGA		5'		
E	5'	AAAAATACGTCGATAG	3'	5'	AGCTCGGGCTA		3'
	3'	ATGCAGCTATCAAAAA	5'	3'	GCCCGATTCGA		5'
F	5'	TACGTCGATAG	3'	5'	CGGGCTA		3'
	3'	ATGCAGCTATC	5'	3'	GCCCGAT		5'

Molecules of the F type can be generated by a process called
random shearing. Since DNA molecules are long rigid rod-type
structures, they tend to break spontaneously when passed through
small apertures. If these types of molecules could be converted
to those shown in A, then reannealing could take place.

In the early 1960s Bollum and his coworkers were attempting to
isolate a DNA polymerizing enzyme from calf thymus. In their
preparations they found an activity which unlike DNA polymerase
could add nucleotides onto the ends of molecules without the
presence of a template. This enzyme, called terminal transferase,
was capable of adding either A's or T's to the ends of molecules
(Yoneda and Bollum, 1965). By controlling the reaction poly A or
poly T tails of varying length could be added to any molecule of

choice. Hence a method of converting F type molecules to A type
molecules became available. As a sidelight it is interesting to
note that the biological function of this enzyme is still not well
understood.

At about the same time that terminal transferase was discovered,
Arber and Dussoix (1962) demonstrated that certain bacterial
viruses (bacteriophage) grown on one strain of *Escherichia coli*
(E. coli K-12) grew poorly on a different strain of the same organism (*E. coli* B). On the other hand, those few virus particles
which survived were fully capable of infecting *E. coli* B but not
E. coli K-12. Hence *E. coli* apparently had a mechanism by which
it could protect itself from foreign DNA. The inability of the
bacteriophage to grow resulted from specific degradation (restriction) of the infecting viral DNA. The enzyme which carried out
this process (restriction endonuclease) introduced a limited number of breaks in the phosphodiester backbone of the DNA molecule
(Figure 1, Figure 3). DNA could be protected from these enzymes
by specifically altering certain nucleotides (modification).
Mutants of *E. coli* which lacked these enzymes were subsequently
isolated (Boyer, 1964; Colson *et al.*, 1965; Lederberg, 1966; Wood,
1966).

Mertz and Davis (1972) provided evidence that the RI restriction enzyme from *E. coli* introduced staggered cleavages thereby
generating overlapping ends that could reassociate through pairing
of the nucleotide bases (Type C molecules). Hedpath *et al.* (1972)
showed that the RI enzyme recognized a symmetrical six nucleotide
sequence, generating cohesive AATT ends (Figure 2). The result
obtained with the *E. coli* RI restriction endonuclease turned out
not to be unique. Many additional enzymes have been isolated from
other bacterial species. They recognize sequences of from 4 to 8
nucleotides, introducing breaks in the phosphodiester backbone
either exactly opposite each other or in some form of staggered
cleavage. (See Figure 2 for three examples.)

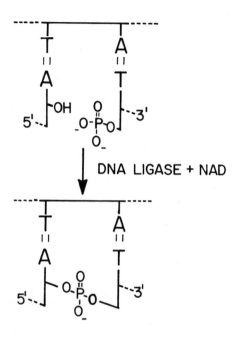

DNA LIGASE + NAD

Figure 3.   The Formation of a Phosphodiester Bond by Polynucleo-
             tide Ligase. This figure portrays the formation of a
             phosphodiester bond by the *E. coli* DNA ligase. NAD
             (a small helper molecule) provides the energy to drive
             the reaction.

D. *Sealing Together DNA Fragments (Ligation)*
     Joining together fragments once they have reannealed requires
an enzyme which can form phosphodiester linkages in the backbone
of the DNA molecule (Figure 3). Such a class of enzymes, called
polynucleotide ligase, was discovered independently in 1967 by
several laboratories (Gellert, 1967; Olivera and Lehman, 1967;
Weiss and Richardson, 1967, Gefter *et al.*, 1967; Cozzarelli *et al.*,
1967). This type of enzyme was shown to catalyze the reaction
depicted in Figure 3, and provided the basis by which covalently
closed DNA molecules could be generated after reannealing had
taken place. This type of enzyme in fact appears to be essential

for cell viability through its involvement in DNA replication and
DNA repair (Gottesman *et al.*, 1973; Konrad *et al.*, 1973).

E. *Replication and Maintenance of Recombinant DNA Molecules*

Even with the ability to splice fragments of DNA together, if
the recombinant molecules could not be maintained in a recipient,
nothing would be gained. Maintenance requires that the DNA be
replicated inside the cell so that at the time of cell division
both progeny contain at least one molecule. Replication of DNA
normally starts at unique sites, such that the entire *E. coli*
chromosome contains a single replication origin. Thus any given
DNA fragment will most likely not contain its own origin of
replication. Accordingly, in order to isolate a particular fragment of DNA, some sort of helper (vector or vehicle) would most
likely be needed for its replication and maintenance. Two types
(vectors) were developed employing information already available
on DNA molecules which could replicate autonomously (plasmids)
and on the bacterial virus lambda.

The first type of vector is based on a class of self-
replicating plasmids whose presence in cells was first indicated
in 1953 (Fredericq and Betz-Bareau, 1953). They presented evidence that a bacterial antibiotic called colicin E1 was carried
on an extrachromosomal element. Some years later Roth and
Helinski (1967) showed that the plasmid, called ColE1, existed as
a covalently closed molecule inside the cell. This particular
plasmid was shown to exist in multiple copies per cell (Helinski
and Clewell, 1972), to contain a single recognition site for the
restriction endonuclease RI, and to be relatively small in size
(Bazarel and Helinski, 1968). In addition, under appropriate
conditions the number of copies per cell could be increased
(Clewell, 1972; Clewell and Helinski, 1969).

Other self-replicating plasmids, physically much larger than
ColE1, were shown to carry resistances to certain antibiotics
(Watanabe, 1963). These larger plasmids normally exist in only

1-2 copies per cell. Cohen and Chang showed in 1973 that a small fragment could be generated from a larger drug resistance factor which still could replicate autonomously in *E. coli*. The major difference between this plasmid (pSC101) and the ColEl vectors described above had to do with the number of copies that existed in each cell and the means by which they were replicated.

Alternatively, if the recombinant DNA could be inserted into the host chromosome, it would be replicated as part of the chromosome. The bacteriophage lambda appeared as a unique candidate to carry out such a reaction. Upon infection the virus can be incorporated in the chromosomal DNA (lysogenized) and replicated as part of the host's genetic material. Under certain circumstances, it could be induced and more virus particles would be produced (Lwoff correctly described this process in his review published in 1953). Also regions of the lambda DNA molecule could be deleted without affecting its ability to replicate. Additionally in order to produce a viable phage particle, the DNA molecule incorporated into the virus particle had to be of a certain size.

Thomas *et al.* (1974) and Murray and Murray (1974) took advantage of these properties to generate special forms of bacteriophage lambda into which exogenous DNA could be introduced. These phages (lambda generalized transducing phages, λgt) contained a limited number of restriction endonuclease sites and a large deletion. Without the insertion of additional DNA these viruses could not propagate in an *E. coli* host. Once recombinant DNA had been inserted, the virus would produce viable progeny.

The location of these two types of vectors within the cell are shown schematically in Figure 4. Each system has certain advantages and disadvantages. The virus vectors are somewhat limited as to the amount of foreign DNA that can be inserted into them and the variety of restriction endonuclease sites they contain. On the other hand, they have the advantage of normally existing in a single copy (Figure 4), which can be induced under

appropriate circumstances to make additional copies. Plasmid
vectors can accommodate larger fragments and have more restriction
endonuclease sites. The ColEl related vectors have the potential
disadvantage of existing in multiple copies within the cell
(Figure 4). Thus if the cloned DNA encodes some product that in
large quantities is detrimental to the host, those cells containing that particular recombinant plasmid most likely will not
survive.

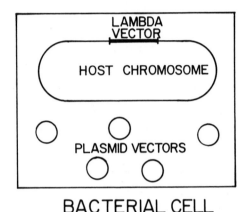

Figure 4.   The Location of Lambda and Plasmid Vectors within the
             Bacterial Cell. The lambda vector is incorporated
             (lysogenized) into the bacterial chromosome and repli-
             cates as part of the host. Under certain conditions
             the virus can be induced and additional copies made.
             Plasmid vectors exist as autonomously replicating DNA
             molecules and occur in anywhere from 1-100 copies per
             cell under normal conditions.

F.   *Introduction of Recombinant DNA Molecules into the Host*
     The introduction of purified DNA molecules into the host
theoretically is not a major problem. Avery *et al.* (1944) had
succeeded in doing this in order to prove that DNA carried biological specificity. Many strains of bacteria could take up

purified DNA efficiently. However, *E. coli* was not one of them.
Finally after much effort, Mandel and Higa (1970) showed that if
E. coli were treated with calcium followed by a short treatment
at elevated temperature, lambda DNA molecules could be taken up
by the cell. Cohen *et al.* (1972) extended this result to show
that plasmid DNA could also be taken up by cells which had under-
gone calcium treatment. Hence another stumbling block was
removed.

G. *Selection of Cells Containing Recombinant DNA Molecules*

The final prerequisite involves the ability to select for
recombinant DNA molecules. In the case of the lambda vectors,
the ability to form viable phage particles was a reliable method
of testing for recombinants. With ColE1 derivatives several
selection methods were possible. Strains harboring ColE1 are
immune to the bacterial antibiotic (colicin E1) encoded by the
plasmid. Accordingly, after a cell has been transformed with
ColE1, the culture can be treated with colicin E1 and survivors
screened for the presence of plasmid DNA (Clark and Carbon,
1975). This method, however, does not distinguish between recombinant plasmids and simple vector molecules. In order to circumvent this difficulty, the phenomenon of antibiotic resistance
has been utilized. Many plasmid derivatives have been constructed
which carry resistance to various antibiotics. Accordingly, the
presence of these plasmids can be detected by the host's
ability to grow in the presence of kanamycin, tetracycline or
ampicillin. Additional means can be used to rapidly determine
if the plasmid contains cloned DNA. Thus certain plasmid
vehicles carry resistance to two antibiotics (Figure 5). Use of
a restriction endonuclease which cleaves within one of the antibiotic resistance genes will lead to inactivation of the gene, if
extra DNA is inserted at the site (the extra DNA will change the
order of the nucleotides thereby altering the genetic code, see
earlier discussion and Figure 2). Transformed host cells which

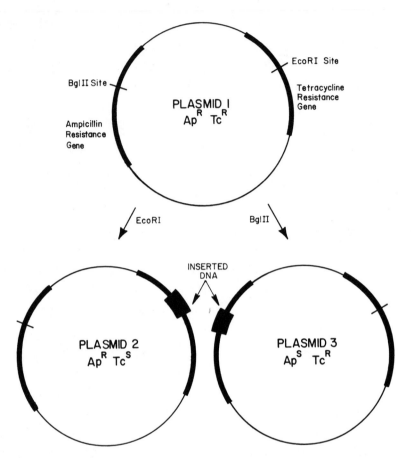

Figure 5.   A Plasmid Vector Carrying Multiple Drug Resistance.
            Plasmid 1 carries resistance to the antibiotics
            ampicillin and tetracycline. Thus a bacterial cell
            harboring this plasmid will be resistant to both drugs.
            Within the tetracycline gene is a recognition sequence
            for the EcoRI restriction endonuclease. If Plasmid 1
            is cleaved with this enzyme and a piece of additional
            DNA inserted at that site, the tetracycline resistance
            will be lost. Hence a cell containing plasmid 2 will
            be ampicillin resistant but tetracycline sensitive.
            The reverse will happen if the Plasmid 1 is digested
            with BglII (Plasmid 3).

are resistant to only one of the two plasmid carried antibiotic resistances most likely contain cloned DNA (Figure 5).

III. UTILIZATION OF RECOMBINANT DNA TECHNOLOGY

Thus between 1972 and 1974 several laboratories using procedures developed in the areas of plasmid biology, bacteriophage lambda, nucleic acid enzymology, DNA replication, drug resistance, transformation and bacterial genetics initiated what is now called recombinant DNA technology (Jackson *et al.*, 1972; Lobban and Kaiser, 1973; Cohen *et al.*, 1973; Morrow *et al.*, 1974; Herschfield *et al.*, 1974; Thomas *et al.*, 1974). Since then the field has progressed rapidly with the development of new vectors, better transformation procedures and new selection methods.

Figure 6 portrays a hypothetical recombinant DNA experiment. In this case a plasmid vector was chosen and cleaved with a restriction endonuclease that leaves small cohesive ends. The DNA from the organism to be cloned is also cleaved with the same enzyme and the two samples mixed together and held at a temperature where the cohesive ends can associate. Following this step, the mixture is treated with DNA ligase to permanently join together any molecules which have annealed. A variety of products is generated by this procedure. The original plasmid vector can be reformed. Fragments of the DNA to be cloned can also form covalently closed circles. These molecules, however, will normally be unproductive because they will not be able to replicate in the bacterial host. Finally, fragments of the foreign DNA can anneal with the plasmid vector to form recombinant or chimeric plasmids.

This mixture of DNA molecules is then used to transform an *E. coli* recipient that will not degrade the foreign DNA (restriction deficient, see earlier discussion of this phenomenon). The presence of the plasmid can be detected by resistance to a

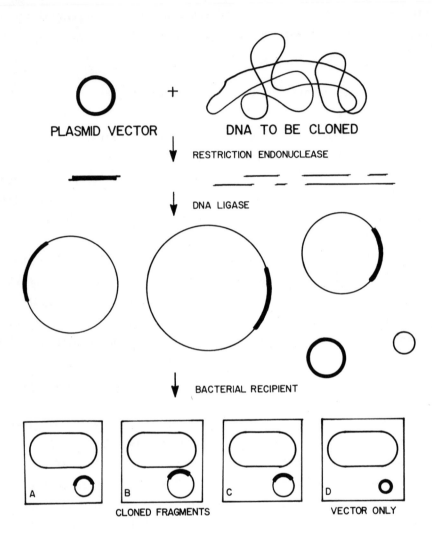

Figure 6.   A Hypothetical Cloning Experiment. This experiment is
            carried out as described in the text. The presence of
            recombinant plasmids can be detected by using some of
            the methods discussed above.

particular antibiotic which is carried on the vector molecule.
The recombinant plasmids (Figure 6A-C) can then be distinguished
from the vector (Figure 6D) by a variety of methods. As described
above (Figure 5), certain plasmids carry resistance to two antibiotics. Thus foreign DNA can be inserted into the structural
gene for one of the resistances. Accordingly after the recipient
has been transformed with the mixture of DNA, transformants
carrying one of the drug resistances but not the other can be
assumed to contain extra DNA.

Alternatively, the presence of cloned DNA can be screened for
directly by the ability of the recombinant plasmid to replace a
missing host function. Thus for example, the host may be unable
to grow in the absence of a particular amino acid because it
cannot synthesize the compound. If the cloned DNA provides the
necessary information to allow synthesis of this compound then
the cell can grow in its absence. This selection method is very
powerful, but requires that the genetic information contained on
the cloned DNA be functionally expressed in the host (Clark and
Carbon, 1976; Vapnek *et al.*, 1976).

Another method of selection depends on the complementary
structure of DNA. The cloned DNA should reassociate (hybridize)
with DNA from the organisms it was derived from but not with the
host DNA. Under appropriate circumstances it is possible to
carry out this type of hybridization experiment immediately
following initial selection. In this manner the presence of
foreign DNA is detectable even if it is not functionally
expressed in the host (Grunstein and Hogness, 1975).

Although Figure 6 describes the use of a plasmid vector and
restriction endonucleases, successful cloning experiments have
been carried out using the other vectors and enzymes described
above (Ratzkin and Carbon; 1977; Struhl and Davis, 1977). To
date a large number of DNA fragments have been cloned and the
list is rapidly growing. Gene products from yeast and certain
fungi have been synthesized (functionally expressed) in *E. coli*

(Ratzkin and Carbon, 1977; Vapnek *et al.*, 1977; Struhl and Davis, 1977). DNA from a variety of organisms has been cloned where functional expression has not been obtained (Kedes *et al.*, 1977; Bedbrook *et al.*, 1977). Some fragments of cloned DNA appear to be unstable in bacterial hosts. A great deal of interesting information regarding the organization of more complicated chromosomes has already been derived from these initial experiments.

IV. DISCUSSION

By using recombinant DNA techniques, scientists will be able to isolate and study genes from complex organisms in ways which have not been previously possible. Such experiments may eventually lead to a greater understanding of many inborn genetic disorders. It may also be possible to have *E. coli* synthesize large quantities of biologically important compounds such as insulin.

Recombinant DNA techniques will speed research in certain areas, but it is not clear that they will permit increased crop production or aid substantially in the fight against disease. Similarly the much discussed potential risks of these experiments also seem more hypothetical than real. It appears unlikely that researchers will create some horrible monster bacteria from these types of experiments, science fiction writers notwithstanding.

On the other hand, the development of recombinant DNA technology has helped bring some basic ethical questions into perspective. Scientists in this century have already made tremendous strides in understanding the processes of living cells. The use of recombinant DNA techniques may speed our ultimate understanding of the mechanisms by which life is created and maintained. Once this occurs it is only a matter of time until these processes can be manipulated. It is at this stage where science and technology can be abused if so desired.

The question then becomes, should science be retarded so as to prevent its possible abuse, or should researchers attempt to move the frontiers of knowledge forward and try to prevent future misuse of the information acquired. History tends to support the concept that the inherent quest for knowledge cannot be surpressed. Therefore, the idea of stifling scientific inquiry does not appear to be realistic. On the other hand, history also provides numerous examples of the misuse of new technology either through ignorance of its long range effects or through political and economic expediency. Accordingly, I would argue that one has to hope that with increased knowledge society will learn to place a greater value on the individual human life and will be able to prevent the abuse of newly acquired knowledge.

ACKNOWLEDGMENTS

This work was supported in part by grants from the National Institutes of Health, GM21454 and GM00048. Ms. Beth Bowen provided invaluable help in the preparation of the manuscript.

REFERENCES

Arber, W., and Dussoix, D. 1962. "Host Specificity of Deoxy-ribonucleic Acid (DNA) Produced by *Escherichia coli*, I. Host-controlled Modification of Bacteriophage λ." *Journal of Molecular Biology*, 5, 18-36.

Avery, O. T.; MacLeod, C. M.; and McCarty, M. 1944. "Studies on the Chemical Nature of the Substance Inducing Transformation of Pneumococcal Types." *Journal of Experimental Medicine*, 79, 137-158.

Bachmann, B. J.; Low, K. B.; and Taylor, A. L. 1976. "Recali-brated Linkage Map of *Escherichia coli* K-12." *Bacteriological Reviews*, 40, 116-167.

Bazaral, M., and Helinski, D. R.   1968.   "Circular DNA Forms of
    Colicinogenic   Factors El, E2, and E3 from *Escherichia coli*."
    Journal of Molecular Biology, *36*, 185-194.

Bedbrook, J. R.; Kolodner, R.; and Bogorad, L.   1977.   "Zea Mays
    Chloroplast Ribosomal RNA Genes Are Part of a 22,000 Base Pair
    Inverted Repeat." *Cell*, *11*, 739-749.

Bolivar, F.; Rodriguez, R. L.; Betlach, M. C.; and Boyer, H. W.
    1977.   "Construction and Characterization of New Cloning
    Vehicles, I. Ampicillin-Resistant Derivatives of the Plasmid,
    pMB 9." *Gene*, *2*, 75-93.

Boyer, H.   1964.   "Genetic Control of Restriction and Modification
    in *Escherichia coli*." *Journal of Bacteriology*, *88*, 1652-1660.

Brenner, S.; Jacob, F.; and Meselson, M.   1961.   "An Unstable
    Intermediate Carrying Information from Genes to Ribosomes
    for Protein Synthesis." *Nature*, *190*, 576-581.

Chargoff, E.   1950.   "Chemical Specificity of Nucleic Acids and
    Mechanism of their Enzymatic Degradation." *Experientia*, *6*,
    201-209.

Clarke, L., and Carbon, J.   1975.   "Biochemical Construction and
    Selection of Hybrid Plasmids Containing Specific Segments of
    the *Escherichia coli* Genome." *Proceedings of the National
    Academy of Sciences USA*, *72*, 4361-4365.

Clarke, L., and Carbon, J.   1976.   "A Colony Bank Containing Syn-
    thetic ColEl Hybrid Plasmids Representative of the Entire *E.
    coli* Genome." *Cell*, *9*, 91-99.

Clewell, D. B.   1972.   Nature of ColEl Plasmid Replication in
    Escherichia coli in the Presence of Chloramphenicol."
    Journal of Bacteriology, *110*, 667-676.

Clewell, D. B., and Helinski, D. R.   1969.   "Supercoiled Circula-
    tion DNA-Protein Complex in *Escherichia coli:* Purification and
    Induced Conversion to an Open Circular DNA Form." *Proceedings
    of the National Academy of Sciences USA*, *62*, 1159-1166.

Cohen, S. N., and Chang, A. C. Y.   1973.   "Recircularization and
    Autonomous Replication of a Sheared R-factor DNA Segment in
    Escherichia coli Transformants." *Proceedings of the National
    Academy of Sciences USA*, *70*, 1293-1297.

Cohen, S. N.; Chang, A. C. Y.; Boyer, H. W.; and Helling, R. B.
    1973.   "Construction of Biologically Functional Bacterial
    Plasmids *In Vitro*." *Proceedings of the National Academy of
    Sciences USA*, *70*, 3240-3244.

Cohen, S. N.; Chang, A. C. Y.; and Hsu, L. 1972. "Nonchromo-
    somal Antibiotic Resistance in Bacteria: Genetic Transforma-
    tions of *E. coli* by R-factor DNA." *Proceedings of the
    National Academy of Sciences USA, 69,* 2110-2114.

Colson, C.; Glover, S. W.; Symonds, N.; and Stacey, K. 1965.
    "The Location of the Genes for Host-controlled Modification
    and Restriction in *Escherichia coli* K-12." *Genetics, 52,*
    1043-1050.

Cozzarelli, N. R.; Melechen, N. E.; Jovin, T. M.; and Kornberg, A.
    1967. "Polynucleotide Cellulose as a Substrate for a Poly-
    nucleotide Ligase Induced by Phase T4." *Biochemical and Bio-
    physical Research Communications, 28,* 578-586.

Crick, F. H. C. 1958. "Protein Synthesis." *Symposia-Society for
    Experimental Biology, 12,* 138-163.

Franklin, R. E., and Gosling, R. G. 1953. "Molecular Configura-
    tion in Sodium Thymonucleate." *Nature, 171,* 740-741.

Fredericq, P.,and Betz-Bareau, M. 1953. "Transfert génétique de
    la propriété colicinogene chez *Escherichia coli.*" *Compte
    Rendues Société Biologique, 147,* 1110-1112.

Gefter, M. L.; Becker, A.;and Hurwitz, J. 1967. "The Enzymatic
    Repair of DNA, I. Formation of Circular λDNA." *Proceedings
    of the National Academy of Sciences USA, 58,* 240-247.

Gellert, M. 1967. "Formation of Covalent Circles of Lambda DNA
    by *E. coli* Extracts." *Proceedings of the National Academy of
    Sciences USA, 57,* 148-155.

Glover, D. M.; White, R. L.; Finnegan, D. J.;and Hogness, D. S.
    1975. "Characterization of Six Cloned DNAs from *Drosophila
    melanogaster,* Including One that Contains the Genes for rDNA."
    Cell, 5, 149-157.

Gottesman, M. M.; Hicks, M. L.;and Gellert, M. 1973. "Genetics
    and Function of DNA Ligase in *Escherichia coli.*" *Journal of
    Molecular Biology, 77,* 531-547.

Grunstein, M., and Hogness, D. S. 1975. "Colony Hybridization: A
    Method for the Isolation of Cloned DNAs that Contain a
    Specific Gene." *Proceedings of the National Academy of
    Sciences USA, 72,* 3961-3965.

Hedgpeth, J.; Goodman, H. M.; and Boyer, H. W. 1972. "DNA
    Nucleotide Sequence Restricted by the RI Endonuclease."
    Proceedings of the National Academy of Sciences USA, 69,
    3448-3452.

Helinski, D. R., and Clewell, D. B. 1971. "Circular DNA." *Annual Review of Biochemistry, 40,* 899-942.

Hershey, A. D., and Chase, M. 1952. "Independent Functions of Viral Protein and Nucleic Acid in Growth of Bacteriophage." *Journal of General Physiology, 36,* 39-56.

Hershfield, V.; Boyer, H. W.; Yanofsky, C.; Lovett, M. A.; and Helinski, D. R. 1974. "Plasmid ColE1 as a Molecular Vehicle for Cloning and Amplification of DNA." *Proceedings of the National Academy of Sciences USA, 71,* 3455-3459.

Hotchkiss, R. D. 1949. "Chemical Studies on the Transforming Factor of Pneumococci." Colloques internationale centre nationale recherche science, Unites biologie douées contin. *Génétiques, 8,* 57-65.

Jackson, D. A.; Symons, R. H.; and Berg, P. 1972. "Biochemical Method for Inserting New Genetic Information into DNA of Simian Virus 40: Circular SV40 DNA Molecules Containing Lambda Phage Genes and the Galactose Operon of *Escherichia coli.*" *Proceedings of the National Academy of Sciences USA, 69,* 2904-2909.

Jacob, F., and Monod, J. 1961. "Genetic Regulatory Mechanisms in the Synthesis of Proteins." *Journal of Molecular Biology, 3,* 318-356.

Kedes, L. H.; Cohn, R. H.; Lowry, J. C.; Chang, A. C. Y.; and Cohen, S. N. 1975. "The Organization of Sea Urchin Histone Genes." *Cell, 6,* 359-369.

Kelly, T. J., and Smith, H. O. 1970. "A Restriction Enzyme from Hemophilus Influenzae, II. Base Sequence of the Recognition Site." *Journal of Molecular Biology, 51,* 393-409.

Konrad, E. B.; Modrich, P.; and Lehman, I. R. 1973. "Genetic and Enzymatic Characterization of a Conditional Lethal Mutant of *Escherichia coli* K12 with a Temperature-sensitive DNA Ligase." *Journal of Molecular Biology, 77,* 519-529.

Kornberg, A.; Lehman, I. R.; Bessman, M. J.; and Sims, E. S. 1956. "Enzymic Synthesis of Deoxyribonucleic Acid." *Biochemica Biophysica Acta, 21,* 197-198.

Lederberg, S. 1965. "Host-Controlled Restriction and Modification of Deoxyribonucleic Acid in *Escherichia coli.*" *Virology, 27,* 378-387.

Lwoff, A. 1953. "Lysogeny." *Bacteriological Reviews, 17,* 269-337.

Mandel, M., and Higa, A. 1970. "Calcium-dependent Bacteriophage DNA Infection." *Journal of Molecular Biology, 53,* 159-162.

Marmur, J. 1961. "A Procedure for the Isolation of Deoxyribonucleic Acid from Micro-organisms." *Journal of Molecular Biology 3,* 208-218.

Mertz, J. E., and Davis, R. W. 1972. "Cleavage of DNA by RI Restriction Endonuclease Generates Cohesive Ends." *Proceedings of the National Academy of Sciences USA, 69,* 3370-3374.

Meselson, M., and Stahl, F. W. 1958. "The Replication of DNA in *Escherichia coli.*" *Proceedings of the National Academy of Sciences USA, 44,* 671-682.

Morrow, J. F.; Cohen, S. N.; Chang, A. C. Y.; Boyer, H. W.; Goodman, H. M.; and Helling, R. B. 1974. "Replication and Transcription of Eukaryotic DNA in *E. coli.*" *Proceedings of the National Academy of Sciences USA, 71,* 1743-1747.

Murray, N. W., and Murray, K. 1974. "Manipulation of Restriction Targets in Phage λ to Form Receptor Chromosomes for DNA Fragments." *Nature, 251,* 476-481.

Nirenberg, N. W., and Matthaei, H. J. 1961. "The Dependence of Cell-free Protein Synthesis in *E. coli* upon Naturally Occurring or Synthetic Polyribonucleotides." *Proceedings of the National Academy of Sciences USA, 47,* 1588-1602.

Olivera, B. M., and Lehman, I. R. 1967. "Linkage of Polynucleotides through Phosphodiester Bonds by an Enzyme from *Escherichia coli.*" *Proceedings of the National Academy of Sciences USA, 57,* 1426-1433.

Ratzkin, B., and Carbon, J. 1977. "Functional Expression of Cloned Yeast DNA in *Escherichia coli.*" *Proceedings of the National Academy of Sciences USA, 74,* 487-491.

Roth, T. F., and Helinski, D. R. 1967. "Evidence for Circular DNA Forms of a Bacterial Plasmid." *Proceedings of the National Academy of Sciences USA, 58,* 650-657.

Struhl, H., and Davis, R. W. 1977. "Production of a Functional Eukaryotic Enzyme in *Escherichia coli:* Cloning and Expression of the Yeast Structured Gene for Imidazole-glycerolphosphate Dehydratase (his3)." *Proceedings of the National Academy of Sciences USA, 74,* 5255-5259.

Taylor, A. L., and Thomas, M. S. 1964. "The Genetic Map of *Escherichia coli* K-12." *Genetics 50,* 659-677.

Taylor, J. H.   1965.   *Selected Papers on Molecular Genetics.*
New York: Academic Press, p. 153.

Thomas, M.; Cameron, J. R.; and Davis, R. W.   1974.   "Viable Molecular Hybrids of Bacteriophage Lambda and Eukaryotic DNA."
Proceedings of the National Academy of Sciences USA, 71, 4579-
4583.

Watanabe, T.   1963.   "Infective Heredity of Multiple Drug Resistance in Bacteria." *Bacteriological Reviews, 27,* 87-115.

Watson, J. D., and Crick, F. H. C.   1953a.   "Molecular Structure
of Nucleic Acids. A Structure for Deoxyribose Nucleic Acid."
Nature, 171, 737-738.

Watson, J. D., and Crick, F. H. C.   1953b.   "Genetical Implications of the Structure of Deoxyribonucleic Acid." *Nature,
171,* 964-967.

Weiss, B., and Richardson, C. C.   1967.   "Enzymatic Breakage and
Joining of Deoxyribonucleic Acid, I. Repair of Single-
strand Breaks in DNA by an Enzyme System from *Escherichia
coli* Infected with T4 Bacteriophage." *Proceedings of the
National Academy of Sciences USA, 57,* 1021-1028.

Wilkins, M. F. H.; Stokes, A. R.; and Wilson, R. H.   1953.
"Molecular Structure of Deoxypentose Nucleic Acids." *Nature,
171,* 738-740.

Wood, W. B.   1966.   "Host Specificity of DNA Produced by
Escherichia coli: Bacterial Mutations Affecting the Restriction and Modification of DNA." *Journal of Molecular Biology,
16,* 118-133.

Vapnek, D.; Alton, N. K.; Bassett, C. L.; and Kushner, S. R.   1976.
"Amplification in *Escherichia coli* of Enzymes Involved in
Genetic Recombination: Construction of Hybrid ColEl Plasmids
Carrying the Structural Gene for Exonuclease I." *Proceedings
of the National Academy of Sciences USA, 73,* 3492-3496.

Vapnek, D.; Hautala, J. A.; Jacobson, J. W.; Giles, N. H.; and
Kushner, S. R.   1977.   "Expression in *Escherichia coli* K-12
of the Structural Gene for Catabolic Dehydroquinase of
Neurospora crassa." *Proceedings of the National Academy of
Sciences USA, 74,* 3508-3512.

Yoneda, M., and Bollum, F. J.   1965.   "Deoxynucleotide-polymerizing
Enzymes of Calf Thymus Gland, I. Large Scale Purification of
Terminal and Replicative Deoxynucleotidyl Transferase."
Journal of Biological Chemistry, 240, 3385-3391.

Genetic Engineering in Agriculture: Biological Nitrogen Fixation as a Case History

Raymond C. Valentine

Plant Growth Laboratory
University of California
Davis, California

GENETIC ENGINEERING: A MISUNDERSTOOD TERM

Genetic engineering is the application of the science of genetics! As such, genetic engineering is a noble and ancient practice going back as far as man's first simple attempts to mold and breed animals and plants for his domestic use--many thousands of years. There has even been given a 'Nobel Peace Prize' for efforts of genetic engineers pioneering the 'green revolution'. Recombinant DNA, the central theme of this volume, is only one of a battery of methods available and should not be considered as synonymous with genetic engineering.

An example of genetic engineering as an ancient art is illustrated in Fig. 1. The figure depicts how the Indians living on the high plateaus of Mexico intervened to domesticate and steadily improve the culture of maize. The small ear on the left, representative of the earliest varieties of domesticated maize, had a cob only about one inch long!

59

Fig. 1.   Increase in size of maize between c. 5000 BC and c.
1500 AD at Tehuacan, Highlands of Mexico.

THE STORM SUBSIDES

The raging tempest regarding genetic engineering has now
calmed to barely a murmur, all in the period of a year or so. Are
we so fickle or are there now answers to some of the questions
heard over and over. What are the benefits? What are the risks?
I don't think society is so fickle; I think there are beginning
to be some answers. What has happened to calm the storm? The
following points probably say it best:

  • There have been no health hazards or deaths yet attributed
    to genetic engineering.

  • Some major benefits are now closer to realization.

The major benefits are visualized in two areas, both intimately
a part of our lives--medicine and agriculture.

In medicine, the very recent exciting discovery that an animal
gene coding for a hormone is capable of expression in the
bacterial host *E. coli* illustrates the magnitude of the potential
benefits of recombinant DNA research.

The next paragraphs contain the case history of a research
area of major consequences to agriculture--biological nitrogen
fixation.

HARNESSING THE SUN: THE GODDESS OF *Azolla*

Biological nitrogen fixation is the conversion of atmospheric
nitrogen gas, a colorless gas found in great abundance in ordinary
air into 'fixed' nitrogen which is needed for making the vital
proteins of all living creatures. By some strange quirk of evolution only the simplest forms of life on earth, the bacteria and
blue-green algae, have developed this process making man dependent

on these wee creatures for almost all the naturally occurring N-fertilizer found in nature. A unique little plant capable of capturing solar energy for making its own N-fertilizer is described below.

Azolla is a small water fern seen as a green mat floating on the surfaces of fresh water ditches and catch basins in many parts of the world. Travelers from the rice belt of Asia have brought back stories of the domestication of this unusual weed for natural production of N-fertilizer in rice paddies of sections of North Vietnam and China.

According to folklore (and with only slight embellishment by the author) the practice of using *Azolla* as a natural source of fertilizer began more than 300 years ago in a remote village in North Vietnam. Ba Heng, a peasant lady of the region, perplexed her neighbors by habitually bringing in an unusually heavy harvest of rice. Her secret involved the culturing and puddling of *Azolla* into her paddy soil before planting. We now know that *Azolla* behaves as an excellent green manure rich in N. The practice spread until today many farmers of the region use this technique. Some enterprising peasants carried out a business on the side of supplying their neighbor with inoculum of *Azolla* for the spring planting. Maintaining cultures of *Azolla* in active, vegetative form through the long winter months for rapid proliferation as inoculum in the spring turned out to be a tricky business mastered by only a handful of families. One reviewer of the literature of this field dug up the interesting story that the formula for over-wintering of *Azolla* became a tightly held secret traditionally passed on by word of mouth from father to son (only upon the son's marriage). Daughters who might stray from the clan were not permitted to share the formula. The secret formula: keep adding wood ashes (potash) from the fire to the *Azolla* pond throughout the winter months maintaining a suitable level of alkalinity! Today, Ba Heng is honored as the 'Goddess of *Azolla*' during an annual fall festival held at the '*Azolla* pagoda' in North Vietnam.

Is this black magic or does Ba Heng's floating weed actually contribute N-fertilizer to the rice paddy? Indeed, using modern techniques, agronomists D. W. Rains and Steven Talley of the University of California at Davis have succeeded in growing respectable crops of rice using the N-input from *Azolla* species native to California waterways. One of the management schemes for *Azolla* at Davis is illustrated in Fig. 2.

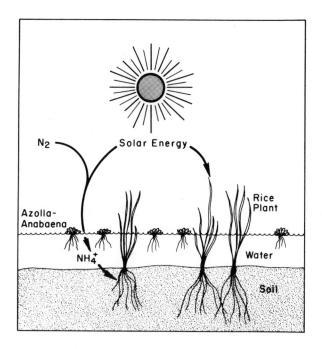

Fig. 2. Harnessing solar energy . . . Biological nitrogen fixation by the water fern *Azolla*.

In this scheme, solar energy is harnessed by the fern in symbiotic association with blue-green algae which reside inside a packet on each leaf. In essence, each leaf maintains its own miniature N-fertilizer factory for supplying its growing cells. During this process, the bountiful gaseous form of nitrogen which makes up the

bulk of our air is broken apart through the action of sunlight and
suitable catalysts yielding a form of N-fertilizer available for
cellular growth. This process is called biological nitrogen
fixation and is governed by a set of genes called the nitrogen
fixation (Nif) genes.

DISCOVERY OF THE NITROGEN FIXATION (Nif) GENES

All processes of the living cell are programmed by the master
hereditary tape of life, DNA. Nitrogen fixation is no exception
and in 1971, a graduate student in the author's laboratory then at
the University of California at Berkeley, discovered and named the
genes for Nitrogen Fixation (Nif). There was not a lot of excitement at the time about this announcement even though the paper
appeared in one of the most prestigious journals in science, the
Proceedings of the National Academy of Sciences, USA. In retrospect, it is easy to understand. This was before the oil embargo,
during an era of great crop surpluses and cheap commercial nitrogen fertilizer. Today, the world is different; we now live in a
climate in which energy is regarded as our number one priority for
research and development. Although the energy used in farming is
only a small part of the whole energy pie, we nevertheless should
be concerned about conservation of energy on the farm. One way
to do this is through enhancement of biological nitrogen fixation
through genetic engineering of the Nif genes. Following closely
on the heels of our work at Berkeley, Ray Dixon, a graduate
student in John Postgate's laboratory in Sussex, England,
succeeded in transferring the Nif genes to the common human gut
microorganism *E. coli* genetically constructing a new hybrid
capable of producing its own N. Although their new genetically
engineered hybrid strain has not yet proven to be practical in
agriculture, the work of the English group has stimulated great

interest throughout the world, pointing to the day when new
varieties of nitrogen fixing organisms may be tailor made.

Although this is an exciting possibility, there are few who
believe that genetic engineering of new N-fixing plants using Nif
genes is very close at hand. The reason for this repressed opti-
mism is summarized on the scorecard in Table I.

TABLE I. *Potential for Genetic Engineering of Nitrogen*
          Fixation

Key ingredients necessary for nitrogen fixation	The colon bacterium, E. coli	Green plants	Animals
Nitrogen fixation (Nif) genes	−**	−	−
High energy electrons (reductant)	+*	+	−
Large quantities of metabolic energy	+	+	+
Protection against oxygen	+	−	−
Regulatory controls	+	−	−

*Present
**Absent

The column on the left hand side of Table I lists the various
essential ingredients (reactions, enzymes, and processes) that
must work together to produce nitrogen fixation. The nitrogenase
genes, though cardinal to this process, are only one of several
sets of genes that must be engineered. For example, the levels
of free oxygen in the cell must be closely regulated or nitrogenase will be destroyed. This is an extremely complicated task for
which we have little understanding yet at the genetic level.

Also, nitrogen fixation requires such huge amounts of cellular energy that special sets of energy genes may be required. In addition, sophisticated genetic control elements are needed to keep a check on cellular activity. In other words, many complicated parts must be "engineered" before new species of N_2-fixing plants can be constructed. The outlook is long range.

Nif AND THE ENVIRONMENT

As we all know, Mother Nature works in wondrous and sometimes strange ways; such is the case with biological nitrogen fixation. Only a relatively small number of higher plants have evolved this power and in all such cases the higher plant is found to associate with a symbiotic microorganism. Working together, the friendly microbe and the host plant cooperate to do a job that is not possible for the plant alone. A question heard often is why haven't all plants including many of our major crop plants such as corn and wheat gained this life-giving ability? Obviously, non-nitrogen fixing plants succeeded through evolution in perfecting their systems for procuring soil nitrates to the point that they were quite able to compete with their nitrogen fixing breathren. This is borne out by the fact that as a general rule, nitrogen fixing species behave as colonizers of poor soil, taking hold and thriving in areas of low soil nitrogen. The advantages of nitrogen fixation disappears after several generations of nitrogen fixing plants have laid down enough nitrogen rich humus allowing non-nitrogen fixing plants to flourish. Some workers now believe that there is some inherent disadvantage borne by plants which make their own N fertilizer. Does the building and running of their own solar nitrogen factory cost a large proportion of the overall resources of the nitrogen-fixing plant? In short, is the process an energy burden that green plants which extract available soil nitrogen don't have to worry about?

There is increasing evidence to back up the point of view that nitrogen fixation is an especially costly process in terms of the overall metabolic energy of the host plant. For example, the root nodules which function as little nitrogen fixing factories on the roots of leguminous plants such as soybean, alfalfa, peanuts, and clover, are believed to be costly to operate, resulting in a lowering of potential yield for these crops. Thus, genetic engineering of nodules on new crop plants may result in reduced yield. (This is of course a mute point in many developing countries of the world and even a few developed countries which either don't have or can't affort to buy commercial fertilizer.)

There may be ways around this problem; for example, several workers are actively studying *Azolla* to see if this organism and its blue-green algae symbiant have succeeded in forming a more energy efficient way of producing available N. It is interesting to speculate that sunlight may be used directly to power the process in the photosynthetically active blue-green algae cells. This differs from the legume system in which sunlight in the leaves is used for producing plant sugars which must travel to the roots as fuel for the nodules, a process in which sunlight is used indirectly with sugar as the 'middle-man' in what amounts to a rather inefficient energy conversion process.

This leads to another important environmental consideration; as already stated, most important biological nitrogen fixation types utilize solar energy, a renewable resource, in contrast to the commercial nitrogen factory which uses natural gas, a fossil fuel feedstock which is in grave danger of depletion in the decades ahead as we near the end of the petroleum age.

The agricultural use of commercial N-fertilizer is projected to increase dramatically in the years ahead, many feel, with the potential for damage to the environment. Since available nitrogen in the soil is known to give rise to "biological smog" (oxides of nitrogen produced in soil and water) through the action of denitrifying bacteria some workers project a fertilizer-ozone

problem in which the stratospheric ozone layer which protects us
from damaging ultra violet rays from the sun may be depleted by
increased N-fertilization. This is a "hot question" which suffers
from insufficient hard scientific facts. However, it is not
difficult to see that biologically fixed nitrogen which is
quickly tied up by the plant and unable to escape into the
environment is potentially cleaner than commercial fertilizer in
which large amounts of added nitrogen is exposed to the action of
soil denitrifying bacteria which produce oxides of N.

LOOKING AHEAD

Mankind is currently consuming energy at a prodigious rate,
particularly fossil fuels. This is a finite energy source which
all too soon will be depleted--the end of the age of petroleum.
Clearly, we must seek new forms of energy with solar energy being
the most attractive. For untold centuries, green plants have
captured and concentrated the sun's rays with their accumulated
biomass serving as a great storage battery of energy for present
and future living things. For example, fossil fuels of today which
power our economy had their origins from the accumulation of an
excess of living biomass, the great oil and coal fields serving
as a reminder of the vast potential of the sun working through the
ages with plant life. What is the ultimate potential of the green
plants as solar energy machines? These frontiers are now being
explored. We have talked here mainly about the harnessing of the
sun for building "solar protein," an essential ingredient in man's
foodchain. The future? Much work has gone into the evaluation of
priorities for future research and development of this area. The
importance of solar energy is illustrated in the following set of
research goals which now guide workers in this field:

1. Develop varieties of leguminous plants such as soybeans which more efficiently convert sunlight into energy-rich photosynthetic products needed for powering symbiotic nitrogen fixation.

2. Genetically engineer strains of symbiotic root nodule bacteria which utilize photosynthetic products more efficiently.

3. Domesticate *Azolla*, blue-green algae, and other plants which use solar energy for producing their own supply of N-fertilizer.

4. Conserve available soil nitrogen through the development of plants and microbes which more efficiently utilize this resource.

5. Genetically engineer new nitrogen fixing crop plants.

These goals are among the most exciting challenges ever faced by mankind. It is the obligation of all scientists involved to provide an open forum with the public concerning all aspects of their work. It is the duty of the public to actively participate in decisions for shaping the destiny of the new genetics.

The Dangers of Unrestricted Research: The Case of Recombinant DNA

Richard Novick

Department of Plasmid Biology
The Public Health Research Institute of
The City of New York, Inc.
New York, New York

I have been asked to write on the dangers of unrestricted research in the area of recombinant DNA. Dr. Sinsheimer has presented a very eloquent philosophical--one might even say transcendental--discourse on this subject, and I now conceive my purpose to be one of dealing with the same subject on a very much more practical plane. This presentation, then, will consider the following two questions:

i. Should man regulate his own scientific and technological activities and, if so, on what grounds should such regulation be undertaken?

ii. Are there biohazards inherent in unrestricted recombinant DNA research of sufficient severity to warrant prior regulation and, if so, how should such regulation be developed and implemented? And, incidentally, are there other areas of modern

71

biological research (such as mutagenesis, cell fusion) that may
also pose a biohazard of sufficient severity to warrant regulation?

With regard to the first question, I have heard the following
argument against regulation: Man is a biological entity and is
therefore subject to the general laws of Darwinian evolution. It
can be argued that therefore whatever we may do to foul our nest is
within the overall evolutionary scheme of things; $i.e.$, if we
pollute ourselves to death or destroy all life on earth by nuclear
warfare or wipe ourselves out by creating a doomsday bug, then, so
be it. The universe will take no notice; other life systems exist
elsewhere or will evolve, possibly suffer a similar fate, etc., so
why be concerned. However, the argument can be extended very
slightly: If whatever self-destructive activities we undertake are
unassailable on Darwinian grounds because we are biological entities, then intellectually conceived actions that we may take to
prevent our own self-destruction are also biological activities and
also subject to the same laws. It may in the long run prove true
that man is a cerebral dinosaur--certainly some human behavior is
consistent with that idea. However, I feel that is no reason to
throw up our hands and say let the fates take us whither they will.
In fact, it is a well established precedent that the activities of
scientists, like other human endeavors, sometimes require regulation in the common interest. In this context, the separation of
basic from applied research is particularly apt, and I will continue to maintain it throughout the discussion.

Applied or technological developments can be dealt with
largely on the basis of risk-benefit analysis. Ideally the society
as a whole and not the manufacturer should decide in every case
whether a particular technological advance is worth the cost in
resources, damage to the environment, etc. With basic research it
is much more difficult to do a meaningful risk-benefit analysis,
since in general, neither the risks nor the benefits will be
readily quantifiable. Consequently, therefore, societies have
not taken the responsibility of closely monitoring basic research

except insofar as deciding how much money should be spent and, in very general terms, how it should be spent. It is only where a very substantial prior risk is perceived that basic research is ever likely to become a matter of direct public concern. In fact, the case of recombinant DNA turns out to be unique in this connection. Yes, the public was and is very concerned about atomic bombs and biological warfare, for example, but in both cases it is not the basic research that causes concern--few objected in the 1940s or would object now to basic research into atomic structure, nor is there any objection to basic research on the biology of pathogenic microorganisms--it is rather the application--misuse, if you will--of basic knowledge that has provoked social reaction.

Accepting the principle that it is appropriate for man to regulate his own scientific and technological activities and to do so for the common good, then I would take the relatively conservative, rather simple position that the overall question of general safety should take precedence over innovation and convenience and that this principle should apply to basic research as well as to technological developments. This, of course, is fairly easy when the danger is clear--not that our behavior often reflects this principle, but at least we know what should be done, for example, about the oil industry and its continual fouling of our oceans, and about the automobile industry and its continual fouling of our air.

What about when the danger is not clear and cannot be evaluated without undertaking the possibly dangerous and probably irreversible activity itself? Shall we interdict in advance on the basis of potential danger? My position here is that such problems can be resolved only by *arduous investigation on a case-by-case basis;* I feel on one hand it was highly appropriate to ban, or attempt to ban a purely technological innovation such as the SST, where the *predictable* harmful consequences seemed heavily to outweigh the advantages--which are no more than mere conveniences. On the other hand, I feel it is inappropriate to attempt to ban a fundamental area of inquiry such as recombinant DNA (R-DNA) research for a variety of reasons:

    i. One can't draw an appropriate line between safe and unsafe and between R-DNA research and other related types of inquiry.

    ii. Danger is only dimly perceived and cannot be adequately evaluated in advance.

    iii. The visible potential advantages seem at the outset to be quite substantial; doubtless many unpredictable advantages exist also.

This is not to say that possibly dangerous R-DNA experiments (and other activities) should be undertaken haphazardly; but rather where hazards are perceived, appropriate precautions must be practiced and constantly adjusted in response to the development of new information.

I. POSSIBLE HAZARDS OF R-DNA RESEARCH

In general, the possible hazards of R-DNA research can be considered in terms of immediate dangers, *i.e.*, the question of safety, and in terms of long-term application, *i.e.*, potential misuse of R-DNA technology. There is, however, far from universal agreement on whether such hazards even exist, let alone on their relative severity. I will shortly discuss the pros and cons of this question, but before that, I would like to discuss briefly one of the possible long-term problems, namely the moral-political question posed by the potential application of R-DNA technology to humans.

A. *The Human Genome: A Moral Issue*

Great concern, which I share, has been voiced over the eventual
possibility of irresponsible tampering with the human genome.
There are, however, two very different principles in this area
which must be considered separately, namely *specific gene therapy*
and *open-ended eugenics*. In order to appreciate this distinction,
one must understand the difference between somatic cells and the
germ line, namely that cells destined to be involved in reproduc-
tion, the germ cells, are differentiated during embryonic growth
from those that go to form the functional organ systems of the
body, the somatic cells. Since modifications of the somatic cells,
genetic or otherwise, are not transferred to the germ cell
line and so are not inherited, one may perform modification of
somatic cells, such as may be envisioned in specific gene therapy,
without having to be concerned about affecting the hereditary
makeup of the individual. Alternatively, in order to engender a
hereditary change, as would be required in artificial eugenics,
one would have to modify the germ cell line.

In specific gene therapy, one would want to isolate and purify
the normal gene, then introduce it into the appropriate somatic
organ of the afflicted individual. In the case of some hereditary
diseases, such as phenylketonuria or hemophilia, this would be the
liver; in the case of others, such as sickle cell anemia, this would
be the bone marrow. As I said, these modifications would not be
hereditary; correction of the germ cell line would be, for technical reasons, rather more difficult and would not relieve the
individual of his affliction. It could be argued that somatic or
phenotypic "gene therapy" would tend to perpetuate disease-
causing genes in the population and so reduce the overall fitness
of the species. I will not get into that argument here except to
say that what is being proposed is no different either in practice
or in principle from the noble art of healing, as has been practiced by physicians since time immemorial.

On the other hand, with open-ended eugenics, the intention
would be to manipulate the genome so as to ameliorate or remove
"bad" traits such as aggressiveness, violence, sexual promiscuity,
etc. or to improve "good" ones, such as competitiveness, manli-
ness, virility, etc. Only one of the fallacies inherent in this
concept is that exemplified by my choice of traits, namely the
impossibility of defining, let alone deciding between, desirable
and undesirable. Another is the absurdity of believing that it
would be possible, even if desirable, to manipulate the genome so
as to obtain the intended result. This is because of technological difficulties inherent in manipulating the germ line and, more
important, because of the impossibility of *ever* obtaining satisfactory information on the genetic basis, if any, of behavioral
traits or of finding out in advance the effect of modifying a gene
such as that involved in the determination of stature, physical
strength, etc. I also find the idea to be grotesquely offensive
on purely moral grounds. Because of these considerations, I would
have no difficulty in supporting specific gene therapy while
rigidly interdicting any intervention in the germ line, including
attempts at correcting specific diseases, as well as open-ended
"eugenic" manipulation even long before it becomes technically
possible. By the same token, I would have no difficulty with the
idea of banning outright the cellular cloning of humans, a technological possibility now facing us. This position is perfectly
consistent with the interdiction on moral and ethical grounds of
various types of experimentation with human subjects simply
because it is humans that are involved. These sorts of bans do
not pose any threat to the advancement of knowledge or to the
alleviation of suffering.

B. *Basic Research: A Safety Issue*

Returning now to the question of whether or not there are biohazards inherent in basic research involving R-DNA techniques, I
would like to outline two opposing arguments, one holding that
there are such hazards and the other holding the opposite.

The analysis usually offered by those who are *concerned* about
the biohazards of R-DNA runs more or less as follows, bearing in
mind that the question of possible hazards can be analyzed both
on the level of safety and on the level of long term potentialities.

1. The first premise is that by artificial DNA recombination
one may generate genetic combinations (novel organisms) that are
extremely unlikely or entirely unable to occur naturally.

2. Although most novel gene combinations will be counteradaptive in terms of survival value, some will surely be adaptive
or at least neutral and therefore potentially able to survive in
the environment.

3. The novel organism, usually a microorganism and therefore
for practical purposes invisible, can multiply, and if it is
viable and should escape from the laboratory, then its fate is
entirely beyond our control.

4. In cases where the novel organism has the potential to
survive in the environment, its accidental or intentional dissemination may have a variety of possibly untoward consequences:
it may have an increased capacity to cause disease, including
resistance to therapy; or it may have an altered ecological potential, or both. By "altered ecological potential" is meant either
the ability to occupy a new habitat or the ability to perform
some new function in a normal habitat, or both. One may

legitimately ask here, what would be so bad about that? The
answer would have to be that we really don't know since it has not,
so far as we know, happened yet. But I can at least point to our
experience with *known* organisms on the assumption that this experience may provide us with a model of expectation. This experience
has been that the introduction of an organism into a new habitat
often has *unpredictable* adverse consequences. As examples, I
might mention water hyacinths, walking catfish, killer bees, fire
ants, chestnut blight, Dutch elm disease, etc. When it comes to
novel recombinants, the same sort of potentials exist, compounded
by an additional degree of uncertainty.

C. Specific Hazards

I have alluded so far to possible hazards of R-DNA in very
general terms. I would like now to outline these in rather more
specific terms, concentrating on the question of safety. The
first possibility I want to mention is that of *untoward consequences of accidental dissemination of novel recombinants.* These may
be entirely unpredictable, resulting from expression of previously
unexpressed genes of unknown function or alteration of genetic
control systems as a consequence of placing genes in a new context. A result of these types of changes could be an entirely
unexpected modification of the biology of the organism into which
the recombinant DNA has been introduced. Other consequences are
more predictable, on the basis of the known properties of the DNA
species involved. For example, bacterial resistance to antibiotics could be modified, jeopardizing the treatment of infections; viruses, viral products, toxins, hormones or other biologically active substances could be produced in a novel context,
or the acquisition of new metabolic activities could affect the
general biology of the recombinant organism. A second possibility, more along the lines of the consequences of application

of R-DNA technology, would be problems raised by the *intentional dissemination of novel recombinants* for some specific purpose. Here, one might encounter unexpected side effects of the *intended* activity of the organism; for example, a novel nitrogen-fixing organism developed for the purpose of enabling one or another of the food crop plants to fix nitrogen might have some unpredictable adverse affect on the nitrogen cycle, or might cause runaway growth of the plant, which would have to be controlled by (toxic) chemicals, etc. Or, alternatively, a recombinant organism released for one purpose, could have other unexpected biological activities; in lieu of an example here, it is important to re-emphasize the general experience that organisms in new habitats often cause trouble in unexpected ways (see below). It should also be pointed out that whereas laboratory strains can be designed to have an extremely low probability of survival in the natural environment, any recombinant organism constructed with the inten-tion of release would obviously have to be hardy.

I would like to make a slightly tangential point here, which is that technological advances in general very often have unpre-dictable consequences in addition to their intended effects. These *unpredictable* side effects are nearly always damaging to the environment or to man or to both, and are responsible for much of our present concern with damage to the environment, cancer causa-tion, etc. Perhaps the simplest examples here are among the chemicals--DDT and other insecticides, aerosols, thalidomide, dyes, etc.--which is not to say that all chemicals have adverse side effects, but rather that when there are unpredicted conse-quences, they are usually negative. In connection with this tangential point, I would like to offer a definition of bio-hazard as *any man-made development, process, substance, etc., that results in an inadvertent modification of the terrestrial bio-sphere,* regardless of whether or not the process or substance in question is in the main beneficial to man or other species.

D. The No-Hazard Position Analyzed

I should like to turn now to the position of those who feel
that there is no biohazard inherent in R-DNA research. Here, the
argument usually runs more or less as follows:

1. Comprehensiveness of Evolution. During the several
billion years of life on earth, all or most possible gene combinations have been tried by evolution through the processes of
mutation and genetic exchange, and any that are not now extant
are counteradaptive. Therefore, it is extremely unlikely that man
could possibly create anything that is (i) novel, or (ii) adaptive for the novel organism. Therefore, there is no significant
biohazard inherent in R-DNA research.

2. Universality of Genetic Exchange. As a corollary of this
argument it has been suggested that although gene combinations may
have evolved to some extent in isolation, genetic exchange is so
widespread, not only among microorganisms, but also between
bacteria and viruses and higher organisms such as man, that any
and all possible advantageous genetic combinations would have
been acquired by any organisms to which they would have been useful. Therefore, any artificial recombinant will be maladaptive.
That is, if free genetic exchange occurs, there is no biohazard
inherent in artificial genetic manipulation, since it has all
happened already.

3. Low Probability of Harm. The probability that a recombinant organism will (i) do anything new, (ii) escape, (iii) establish itself in the natural environment, and (iv) cause significant harm to man or to the general biosphere is already extremely
small and can easily be made much smaller.
Those who use argument number three clearly recognize the
existence of a hazard; the argument simply maintains that the risk

level can be made very low and the only source of disagreement here involves the magnitudes of the numbers used to calculate the various probabilities.

I regard the arguments raised in points one and two, however, as highly spurious on the following grounds and feel that they need to be considered in some detail:

1. *Has Evolution Stopped?* They imply that we just happen to be at that point in time where evolution has essentially stopped, since all or most useful genetic arrangements have already been tried.

2. *Most Gene Combinations Have Yet to be Tried.* On the contrary, one can easily do a calculation to show that only an infinitesimal fraction of all possible genetic combinations could possibly have been tried throughout the course of evolution thus far. This calculation runs as follows:

a. Suppose earth were a solid mass of bacteria containing equal numbers of two unrelated species.

b. Suppose these bacteria have been multiplying every 20 minutes since the earth was formed (4 billion years, more or less).

c. Suppose one genetic exchange between these species took place every time a cell divided (*i.e.*, every 20 minutes).

d. This would give a total of 10^{65} new combinations that would have been generated so far, a number that is obviously a vast overestimate of what could possibly have actually occurred.

e. However, this unfathomable number pales in comparison to what can be generated by combinatorial analysis of possible gene exchanges.

f. If each of the two species has 10,000 genes--a reasonable number for bacteria--and we consider only exchanges of whole

genes and insertions between genes, then there would be 10,000 x
10,000 = 10^8 (10,000,000) single gene exchanges possible between
our two species. However, if we consider more complex genetic
events, for example, exchanges of 50 genes chosen at random, then
there are about 10^{270} different exchange events possible, and this
for only one of a very large number of possible types of exchange
events. For comparison, I have calculated the volume of the
known universe in cubic angstrom units--which is about the largest
number I can generate in physical terms. This turns out to be
10^{114}, assuming a sphere with a radius of 10^{10} light years.

   3. Improbability of Genetic Exchange Among Unrelated Species.
While there may be some truth to the thesis that genetic exchange
among unrelated species can occur, I would argue that this is far
from a general phenomenon but occurs, if at all, only under very
special circumstances. The first of these is in the laboratory
where it can be shown that foreign DNA can be introduced into
cells of various species. In fact, cells that take up DNA make no
distinction in the uptake process between native and foreign DNA.
But just because an organism can be coaxed to take up foreign DNA
in the lab, one has no right to assume that it does so regularly,
if at all, in the wild; and further, even if *uptake* occurs,
experimental evidence now available suggests that incorporation of
foreign DNA into the cell's genome is a very special and unusual
event.

   So far a number of different situations have been described in
which incorporation of foreign DNA has been demonstrated, and it
seems instructive to describe these briefly:

   a. Direct homology of base sequences is the classic basis for
natural recombination; however, it is a general phenomenon only
for very closely related species and even then is usually not
uniform throughout the respective genomes.

b. Conserved genes. Certain genetic traits appear to be
highly conserved throughout evolution, perhaps because only relatively minor variation is consistent with efficient functioning.
The best known example of this is genes determining ribosome
structure in various species of bacilli. Interspecific genetic
exchange involving these genes occurs readily where other genes
can be exchanged either not at all or at an extremely low level.

c. Plasmids and viruses are extrachromosomal genetic systems
that can often inhabit or infect more than one species. Either
may carry genes that affect the host organism's phenotype and may
also incorporate, on occasion, segments of bona fide chromosomal
DNA from one host that they may then transmit to a second. In this
connection, it should be noted that no plasmid or virus that can
cross the eukaryote-prokaryote barrier has yet been encountered
(but see below). However, there are plasmids with extremely wide
host range among the prokaryotes and viruses with extremely wide
host range among the eukaryotes—although the latter are usually
relatively virulent (*e.g.*, yellow fever) and have not been implicated in host gene transfer.

d. Translocatable elements are a class of genetic elements
that have the ability to translocate intact from one genetic site
to another *within the same cell.* They are mentioned in this connection because they provide an added dimension to the interspecific exchange potentiality. A plasmid or a virus may acquire
such an element, may then enter the cells of a second species,
and the translocating element may then become attached to the
genome of the new host *even if the carrier plasmid or virus
should be unable to replicate in that host.* This sequence of
events (involving plasmids) could be responsible for the recent
acquisition of penicillin resistance by *Hemophilus influenzae* and
Neisseria gonorrhea, two species that have hitherto been sensitive
to this antibiotic.

e. Endosymbiosis. A large number of symbiotic relationships between eu- and prokaryotes have been described, such as killer particles in paramecia, algal cells inhabiting and perhaps providing photosynthetic activities for single and multicelled eukaryotes, down to the very organelles of our own cells, and those of higher plants. Organelles are subcellular particles that occur in eukaryotic cells and fulfill major metabolic functions. The most thoroughly studied ones are the mitochondria, which occur in all higher cells and are involved in the metabolic processes by which the cell derives energy from its foodstuffs, and chloroplasts, which occur only in green plants and are the organs of photosynthesis. These two organelles are widely regarded as endosymbiotic descendants of originally free-living prokaryotes; as such they have retained their own genetic material and are to a certain extent hereditarily autonomous within the cells that they inhabit. Here, then, is a prime opportunity for exchange of genes; however, there is as yet no clear evidence on whether or not it has occurred.

f. Other and as yet unimagined possibilities exist. I would place in this category the rather fanciful suggestion that *E. coli* and other bacteria have had ample access to mammalian DNA through their presence during the decay of corpses. Because *E. coli* take up DNA extremely inefficiently, if at all, except under highly artificial conditions; because there is no homology between bacterial and eukaryotic DNA; and because degradative conditions in corpses are almost certainly such that free DNA would survive for no more than a very short time, I would regard the probability of *E. coli* ever having assimilated mammalian DNA under such conditions to be so small as to far outweigh the numbers of corpses and bacteria that have ever existed.

The third and fourth of the above possibilities are of greatest interest here. Theoretically it is possible that any gene *could* be carried by a plasmid, virus, or translocatable element.

So far, in nature, only very special genes have been found
naturally to be carried by such elements; so while I can accept
the *possibility* of exchange between unrelated species, involving
a variety of genes, I have yet to be convinced that it actually
happens except in very special cases.

E. *Genetic Exchange Between Eukaryotes and Prokaryotes*

One of the chief concerns of those who are worried about R-
DNA is that it allows the breaching of a natural barrier between
lower organisms, known as prokaryotes (pro = pre; karyon = nucleus)
because their cells lack a cytologically differentiated nucleus,
and higher organisms known as eukaryotes (eu = normal; karyon =
nucleus) because their cells contain "normal"-looking nuclei. The
prokaryotes include bacteria, viruses, certain algae, rickettsiae,
actinomycetes, chlamydiae and mycoplasmas and as such include
most of the pathogenic microorganisms and those that produce antibiotics. The eukaryotes include all of the higher forms of life
with which we are familiar plus a large variety of microorganisms,
some of which are pathogenic, such as amoebas, trypanosomes,
plasmodia, etc.

Some hold that the prokaryote-eukaryote genetic barrier is
never breached in nature and we would be opening a pandora's box
of potential disasters were we to do so. Others are convinced
that the exchange of DNA is very frequent and the reason that the
genomes of such diverse organisms are generally entirely unrelated, as indeed they are, is that foreign genetic material is
generally of no particular use to the organism and so is dis-
carded.

Although I consider the latter argument as basically spurious
because organisms do not in general possess the elaborate biochemical equipment that would be required to incorporate foreign
DNA unrelated to their own even if it were useful, and because it

is unlikely that the possible utility of foreign DNA could be
realized *unless* it were incorporated, nevertheless, there are a
few cases of apparent or possible genetic exchange between
eukaryotes and prokaryotes, and a word or two of comment on these
might be instructive here.

1. *Viruses and Plasmids.* Although viruses are formally
classified as prokaryotes, their evolutionary and functional
resemblance to cellular prokaryotes such as bacteria is hardly
greater than their resemblance to eukaryotes. Plasmids are not
classified at all but are generally regarded as supernumerary small
chromosomes. My personal view is that for taxonomic purposes,
plasmids and viruses should be regarded as related subcellular life
forms entirely distinct from either prokaryotes or eukaryotes. In
any case, genetic exchange between plasmids and viruses and their
respective host cells is commonplace. In prokaryotes, plasmids may
serve to transport special genes among unrelated species; with one
exception, however, (see below) they have not been implicated in
gene transfer between eukaryotes and prokaryotes. In eukaryotes,
the exchange of genes between virus and host cell is involved in
viral tumorigenesis, and if one insists on regarding viruses as prokaryotes, then viral tumorigenesis can be claimed as an example of
prokaryote-eukaryote exchange.

2. *Crown Gall Tumor* is the most common plant cancer. It is
caused by bacteria of the species *Agrobacterium tumefaciens;* however, once the tumor has been initiated, one can no longer recover
any bacteria from the tumor tissue. It has been known for some
time that tumor-causing strains of *A. tumefaciens* always carry a
large plasmid, and it has very recently been shown by Chilton and
coworkers at the University of Washington (Seattle) that a small,
unique fragment of the plasmid has become integrated into the
chromosomal DNA of the tumor cells (but not of the normal cells of
the same plant). Here again, if one insists in regarding plasmids
as prokaryotes or as parts of prokaryotic genomes, then this can

be claimed as a case of prokaryote-eukaryote genetic exchange, albeit an extremely special one. The formal analogy between this case and that of viral tumorigenesis in animals is rather striking.

3. *"Mammalian" Proteins in Bacteria.* It has recently been observed that certain strains of *E. coli* produce a protein that is very similar or identical to human chorionic gonadotropin, a hormone involved in the maintenance of pregnancy. This observation has given rise to the assumption that because a bacterial protein resembles a mammalian one whose function has no obvious bacterial counterpart, it is assumed that the bacterium must have acquired the gene for that protein from a higher eukaryote. This is based on the further assumption that proteins with seemingly metazoan-specific functions must have evolved *de novo* during or after the differentiation into higher metazoans.

Here is a logical bind: On the one hand, if the exchange assumption is correct, it says that eukaryote genes with no obvious selective advantage may be incorporated by bacteria and persist in nature. If this is so, then the argument that artificially introduced eukaryote genes cannot survive in prokaryotes becomes somewhat tarnished. On the other hand, it may be that the proteins in question are of some use to the bacteria. In that case, why assume they have been acquired from eukaryotes? Why not the reverse, or, even more likely in my view, is the possibility of coancestry. That is, it seems very probable to me that many of the really potent proteins may have evolved functionally before the prokaryote-eukaryote differentiation took place and in the subsequent course of evolution, such ancestral proteins would have been modified in a manner consistent with Darwinian adaptivity for their owners. Structural and even functional similarities could easily have been preserved under such circumstances, so there is no *need* to postulate relatively recent eukaryote-prokaryote exchange to account for these similarities.

While restriction endonuclease EcoRI-mediated insertion of mouse DNA into an *E. coli* plasmid in whole *E. coli* cells (Chang &

Cohen, 1977) demonstrates the *possibility* of incorporation of
eukaryotic DNA into living prokaryotic cells, the conditions under
which this incorporation took place are so extremely artificial
that there is essentially no chance of their occurring in nature;
they do not, in my view, increase at all the likelihood that
eukaryote-prokaryote genetic exchange is a natural phenomenon.

So where does all this leave us? While it is certainly probable
that some genetic exchanges between unrelated organisms have taken
place, I am far from convinced that such exchanges are either general or frequent, and I feel strongly that their mere existence does
not give us *carte blanche* to manipulate with impunity. That is,
even if it is theoretically possible for some particular exchange
to occur, the probabilities are in general negligibly small, and
therefore artificial genetic recombination entails a very real bio-
hazard--and, I might add--so do certain other types of genetic
experimentation such as mutagenesis.

F. *Examples of Possibly Hazardous*
    R-DNA Experiments

In any case, I am quite convinced that it is very possible to
develop a recombinant organism that could cause plenty of trouble,
regardless of whether prior genetic exchanges were common or nonexistent. In order to put this on a more tangible level, I would
like to give as two examples, what one refers to since Watergate as,
scenarios both involving bacterial systems because I believe that
one can make reasonably accurate predictions with such systems:

The first is that a cellulase-producing *E. coli* has been
developed for the purpose of enabling man to digest cellulose fibers
(I believe the organism has been destroyed recently). Presumably
the gene for cellulase would have come from one of the various
cellulose-degrading bacteria. This experiment could have been performed legitimately under the NIH Guidelines.

On developing the organism, the investigator might have wished
to test it for its ability to permit humans to digest cellulose
fibers and so benefit nutritionally. It is very possible that he
might decide to go ahead on the assumption that the gene was probably beneficial and surely harmless. It is conceivable that the
cellulase gene might give the organism an enhanced ability to survive in the human colon--or at least not hinder it. The human
tests would then inevitably be tantamount to releasing the recombinant organism which might then become a widespread inhabitant of
the human colon. Assuming for the moment that the only new biological activity of this organism was that of cellulose digestion, it
would seem worthwhile to ponder some of the possible consequences.

Although, as pointed out by Dr. Formal (this volume), cellulose
digestion in the colon would not enable us to obtain nourishment
from non-nutritive sources since absorption of nutrients does not
occur in the colon, it might nevertheless have possibly disastrous
side effects. The human digestive system has evolved over many eons
to deal with a certain type of natural diet. It has been shown that
the presence of a substantial undigestible residue--mainly of cellulose fibers--is of major importance for the maintenance of normal
bowel function (see D. Burkitt, 1971). Indeed, it has been shown
that our highly refined, low residue diet is probably responsible
for a number of non-malignant but rather unpleasant intestinal disorders such as constipation, irritable colon and ulcerative colitis,
diverticulosis, adenomatous polyps, hemorrhoids, etc., and it has
been suggested that such a diet is also responsible for the
increased frequency of cancer of the colon and rectum that has been
seen in our society in comparison with other groups that eat a more
natural, high residue diet.

While the causal chain here is far from established, the possibility of trouble along these lines is a very real one; it is perfectly possible that a scientist, either ignorant of this possibility or aware of it but deciding that the risk is acceptable,
could go ahead with this line of experimentation and accidentally

release the resulting organism, and it seems to me very likely that somewhere along the line we either will underestimate a possible danger of this sort or fail entirely to perceive it. The number of such possibilities is limited only by the imagination.

Note that in the cellulase case, I can make a plausible prediction of possible adverse consequences only because there happens to be information on the relation of diet to diseases of the colon. This information is relatively new, however, and it is more than likely that other such "beneficial" experiments will be envisioned for which similar information does not happen to be available.

As a second example, let us consider *Clostridium botulinum* (the producer of botulin toxin--the cause of botulism, and the most potent toxin known).

Now, it has often been suggested that one could take the toxin-genes from *C. botulinum* and put them into *E. coli*. However, I can envision a different scenario that might be even more frightening as a biohazard.

As you may know, the *clostridia* are obligate anaerobes, *i.e.*, are able to grow only in the absence of oxygen, and they will not grow in an acidic environment. Consequently, they can contaminate relatively few types of food and, in addition, cannot multiply in the human intestine or in living mammalian tissues. One could conceivably introduce genes permitting growth in the presence of oxygen or in an acidic environment into such an organism, so modifying profoundly its ecological potential. Possible consequences of the ability to grow aerobically would be the ability to infect humans and other animals with dreadful consequences and the ability to multiply in habitats exposed to oxygen. A consequence of the ability of *C. botulinum* to grow in an acid environment would be the lethal contamination of foods (such as pickles, preserved fruits, etc.) that are at present entirely safe. Aside from the obvious public health nightmare that such an organism would represent, one would be changing in a major way

the ecological potential of an important saprophyte. The strictly
environmental consequences of this change would be hard to predict; however, there would surely be such consequences. Finally,
there is the biological warfare potential inherent in such an
organism--which hardly needs to be spelled out. But why could
such experiments not be done under P4 containment? Because even
P4 is not perfect and, especially with an organism that forms
spores, such as *C. botulinum*, the risk of escape by accident,
etc., is simply not acceptable, in my view (and in the view of
the NIH Guidelines).

II. A POSSIBLE COURSE OF ACTION

Granting that all of the foregoing is valid, we have still to
decide on an effective course of action. This course, I believe,
must take the NIH Guidelines as a point of departure, because I
feel the approach taken in the Guidelines is appropriate and
because the Guidelines constitute the most thorough and carefully
throughout-out document available in this area. The basic principle upon which the Guidelines are based is that of evaluating
relative risks and matching them with appropriate precautions. In
this context, there are three classes of experiments:

i. those that are as safe as standard experiments in bacterial genetics and can be performed in the open laboratory;

ii. those that would seem to be unacceptably dangerous and are
not to be performed at all;

iii. those in-between that may be performed only under appropriate containment conditions, including biological as well as physical measures.

By now, most people would agree about the existence of these
three classes of experiments and about how to deal with two of
these classes, the first and second (although there would surely

be considerable disagreement about precisely which experiments
should be placed in these categories). Most of the difficulty
involving the safety of recombinant DNA research revolves around
class iii. experiments--those that have an unpredictable biohazard. Here, there are three general opinions: ban them since
the risk is not worth it; perform them openly and freely since
there really is no hazard; perform them under appropriate containment conditions. And the latter is largely what the NIH Guidelines are all about.

A. *The NIH Guidelines*

Although the Guidelines, as I mentioned, represent the
thoughtful input of many people over a period of nearly three
years, they are still far from perfect, and I should like now to
comment briefly upon some of these imperfections and what might
be done to rectify them.

I see three main problems:

i. It is impossible to assess risk levels accurately or, in
some cases, at all.

ii. It is inherently impossible to achieve a uniform level of
compliance with the Guidelines, and therefore their effectiveness
will inevitably be variable.

iii. The Guidelines have, at present, very limited coverage.

With regard to the first problem, one can do no more than
one's best, and I believe that this has been done, although I
recognize that the results of the analysis of risk are and will
always be open to question.

The second problem, that of compliance and effectiveness,
deserves a more extensive comment. As has been mentioned
repeatedly, the NIH Guidelines were formulated on the basis of
containment; the only consideration given to the question of

performing one or another experiment was whether or not it could
in theory be safely contained.

In considering this principle, there are two questions:

i. Are the Guidelines good enough to ensure safety, given
absolute compliance?

ii. Is absolute compliance likely, or even possible?

I should like to answer both questions with a qualified negative on
the basis of the following considerations, bearing in mind that
proposed containment practices, both physical and biological have
been carefully thought out and will be more or less effective
depending on certain very important qualifications.

B. *Physical Containment*

Considering first physical containment, the only system that
is designed *in principle* to be absolutely effective is that now
referred to as P4. P4 containment, at its best, does not offer
perfect protection--lab accidents and infections have been well-
documented, and despite the existence of P4 facilities, there are
laws on the books forbidding the importation into the U.S. for any
purpose whatsoever, of certain microorganisms. Lower levels of
containment must, of course, be increasingly less effective. Thus,
one has *in principle* a graded series of probabilities with respect
to the escape of organisms. As you know, this has been matched to
what is considered to be a graded series of risks. While this
matching seems, *a priori*, to be reasonable enough, aside from the
unanswerable question of whether or not the risk assessments are
valid, the fact remains that the system is *designed to permit a
higher probability of escape* for organisms regarded as lower on
the risk scale.

In other words, at its very best, the system is not designed
to prevent absolutely the escape of any recombinant organisms

except those perceived *a priori* to be the most dangerous. Moreover, one may with a great degree of certainty assume that in
practice the containment systems will not operate at their optimum
level of effectiveness. Without impugning either the competence
or the responsibility of the practicing scientist, it is difficult
for me to imagine that every laboratory will be able to master and
put into practice all of the procedures recommended by the NIH
Guidelines and described in a 93-page booklet on physical containment (Appendix D of the NIH Guidelines). And, in any case, the
description of these procedures leaves much to the imagination and
the discretion of the individual investigator.

On the one hand, microbiologists who are accustomed to working
with dangerous pathogens will generally have gone through an
elaborate training procedure and will be greatly concerned for
their own safety. Such workers must, in general, be presumed to
be very careful, indeed. Nevertheless, there is a considerable
frequency of laboratory-acquired infections with serious pathogens, despite these precautions.

On the other hand, most of those for whom the NIH Guidelines
are intended are not now trained in microbiological safety techniques. In fact, most of them have chosen to work with *E. coli*
and other relatively innocuous microorganisms *precisely* because
with such organisms one didn't have to worry about contamination,
dissemination and infection. One could, therefore, work with
great rapidity, unhindered by any consideration of safety except
that of protecting one's cultures from gross contamination by
extraneous organisms.

Therefore, it is my belief that even with the best of intentions, it is extremely unlikely that all laboratory heads (who
are the responsible authorities under the NIH Guidelines) will be
able immediately to adopt the full range of necessary safety precautions; and what is more, even if all of the procedures recommended in the Guidelines were carried out to the letter, there
would still be gross breaches of containment inherent in the

system because the Guidelines do have significant deficiencies, and the average biochemist-molecular biologist does not have the training and tradition of safety awareness to perceive these. To be concrete, let me give four simple examples in the area with which I am most familiar, namely handling of bacterial cultures:

1. There is no detailed treatment anywhere in the Guidelines, or in Appendix D, of the aerosol generated in a shaking culture flask and of what to do about it.

2. On page D21, there is a serious misstatement regarding aerosols in connection with the use of a blender: "Before opening the safety blender bowl, permit the blender to rest for at least one minute to allow *settling of the aerosol cloud*." As you may know, an aerosol is an air suspension of very fine particles that may be infectious, are dispersed by air currents, are the major route of inhalation infection, and *do not appreciably settle with time*.

3. A recommendation for opening ampoules (page D15) seems to me both unproven and extremely dangerous:

The researcher uses an intense, but tiny, gas-oxygen flame and heats the tip of the hard glass ampoule until the expanding internal air pressure blows a bubble. After allowing this to cool, he breaks the bubble while holding it in a large low temperature flame; this immediately incinerates any infectious dust which may come from the ampoule when the glass is broken. Preliminary practice with a simulant ampoule of the same type actually in use is necessary to develop a technique that will not cause explosion of the ampoule.

There is no proof that the low-intensity flame is effective. This technique is extremely likely to lead to accidental explosion of ampoules, even in the most careful of hands.

4. Wire innoculation loops are in extremely wide use for transferring cultures, and their use, especially their flame sterilization, has been shown to disseminate infected aerosols. There is no mention of this in the Guidelines, and I would argue that the level of awareness of the problem is very low.

Although there are many similar instances, these four should suffice to make the point.

I think, therefore, that there will be a great deal of variability among laboratories and among individuals in the rigor with which the containment procedures are practiced, and there are two considerations here: First, there are some people who, despite the best of intentions, will have a difficult time observing the required discipline. Second, and perhaps more important, there are some people who believe that all of the present concern over the biohazards of novel biotypes is absurd; there are *no* biohazards. There are a number of rationales for this feeling--which I won't elaborate upon here--but I have very serious doubts that such people would be able, even if willing, to adhere rigorously to a prescribed set of elaborate and rather inconvenient procedures.

In addition, there is the question of natural human error-- which is surely inevitable. I mean such things as spills, mislabelings, various sorts of contamination, mechanical failures of physical systems, etc. In fact, to supervise the operation of a rigorous containment system even as outlined in the Guidelines, for a laboratory of any size, would be an extremely time-consuming and, I might add, boring and unpleasant job.

C. *Biological Containment*

With respect to the other type of containment, namely biological, I would like to make two points:

1. On the Letter and the Spirit of the Guidelines. One laboratory that had been cloning fragments of yeast DNA in an *E. coli* K12 recipient organism that contains a conjugative plasmid has recently destroyed all of the relevant cultures in response to discussions with colleagues. Another laboratory has described the development as a cloning vehicle of a rather dangerous plasmid. This is a conjugative plasmid with the highest known degree of promiscuity (*i.e.*, ability to infect very distantly related organisms). These experiments are certainly prohibited by the Guidelines, although the prohibition is not spelled out precisely as such. The relevant text, under biological containment, reads: "The host is always *E. coli* K12 and the vectors include non-conjugative plasmids and variants of bacteriophage λ." I have the impression that the heads of both of these laboratories consider the Guidelines and all associated regulatory activities to be unnecessary.

2. Safe Organisms. A great deal of effort has been spent recently to develop host organisms and cloning vectors that have a greatly reduced chance of survival outside the laboratory. This effort has been highly successful on paper, but there are a number of potentially serious problems with it as a method of safety.

a. The human factor. The bacterial host, $\chi 1776$, developed by Dr. Roy Curtiss and coworkers, is a very poor grower, is inconvenient to use, and requires constant monitoring to verify its many biological properties. Unless extreme care is exercised in its handling, it is very likely to be replaced in culture by a rapidly growing mutant, or by an extraneous contaminant. I have heard the sentiment voiced that what is wanted is a fail-safe organism so that one can happily forget about physical containment. It is a certainty that $\chi 1776$ is not such a strain; in fact, I seriously doubt that it would be possible to produce one. The

idea of a fail-safe system is an illusory and dangerous miscon-
ception.

    b. Biological problems. Along with the reduced viability of
χ1776 has gone sensitivity to lysis (dissolution) under various conditions, notably the action of bile salts and detergents, as well
as mere growth outside the laboratory. Lysis results in the immediate release of all contained DNA. While there are some situations
in which it is very likely that free DNA will be rapidly degraded,
many others have not been tested, and I can quote at least two different examples of laboratory experiments in volume transformation
that were spoiled by the presence of plasmid DNA as a contaminant.
In both cases this contaminating DNA was taken up and expressed by
the experimental organisms. Finally, there is the possibility that
crippled organisms may find some unsuspected escape route from their
biological straightjackets. For example, the χ1776 strain is
severely compromised in its ability to synthesize the envelope or
wall that protects the bacterial cell. Some bacteria under such
conditions undergo a transition to a stable, wall-less state
referred to as an L-form. These L-forms are morphologically difficult to identify and are thought to be capable of inhabiting the
tissues of a higher organism more or less indefinitely in a form
that is very difficult to detect, identify and culture.

*D. Applicability of the
   Guidelines: Recommendations*

    With respect to the coverage of the Guidelines, this is
limited in two ways, scientifically and politically. Scientifically, the coverage is limited essentially to *E. coli*. I understand that this will be dealt with shortly. Additionally, the
Guidelines cover only R-DNA. For example, one could introduce a
penicillin resistance plasmid into Group A β-hemolytic strepto-
cocci with impunity. This sort of experiment should be covered,

and I personally would like, in addition, to see some attention
given to mutagenesis, cell fusion experiments, and especially to
the dissemination of antibiotics and the consequent selection and
spread of resistant organisms.

If the current picture of the regulatory systems is as bleak
as all that, then where does this leave us? Why do we not insist
on banning all potentially hazardous experiments? One part of the
answer is that while the Guidelines are far from perfect, they
have consistently erred on the side of stringency and caution;
probably if perfection or something near to it could be assured,
the rules could be much less strict. That is, a very ample safety
factor has been built into the regulations themselves. Another
part of the answer is that the hazards of R-DNA research are
generally of a highly speculative nature, and it will be a very
long time before most of them become any less speculative. Therefore, although it is appropriate to take reasonable, perhaps even
overly strict precautions, it seems to me out of line to abandon
a very promising line of research on the basis of imperfect controls for speculative hazards.

At the very least, however, the precautions should be uniform,
consistent, comprehensive and workable, and I would suggest that
the following improvements in the regulatory system would greatly
enhance its effectiveness:

1. Clarification of specific deficiencies and ambiguities in
the Guidelines.

2. Extension of the Guidelines to political and scientific
areas not now covered--hopefully to the entire world, eventually.

3. The institution of proficiency requirements for laboratory
workers in the area. These would be met by training programs.

4. The institution of sanctions: licensing, patents, journal
policies, legal accountability, etc.

What about interference with freedoms: scientific, academic, the press, etc.? The answer goes back to the first question I discussed: society ideally operates by the Millsian doctrine of the greatest common good and unhesitatingly restricts individual freedom where it feels individual actions pose a social threat. And I believe that science has no inherent claim to be exempt from this doctrine.

In conclusion, however, I find myself with a somewhat contradictory position, as follows: Controls are appropriate but are *certain* to be imperfect, yet I believe the overall threat is not great enough to justify interdiction, even if that were possible. Carefully developed controls are the very best that can be done, even if imperfect. We take risks all the time and we are taking a risk here. But let us act so as to minimize the risk, although it is surely impossible to eliminate it.

I would argue, in addition, that the possibly adverse long-term technological consequences of R-DNA research are not of a greater order of magnitude than those of other research activities and can and should be dealt with as the need arises. I do not feel that the danger of such consequences is sufficient in this or in any other area to warrant interdiction of the basic research effort. Therefore, I feel it is appropriate to direct attention to the basic research process and to the problem of dealing with the very real hazard that has been perceived in connection with recombinant DNA research. This hazard is one that transcends the individual scientist and so must be dealt with in a collective manner so as to protect the public at large. I feel that the primary concern for safety that is the basis for all or most of the regulatory activities that have been undertaken is correct.

In this connection, I am in favor of a universal regulatory system that includes legislation, licensing, legal sanctions, and probably most important, training and the achievement of proficiency in the handling of dangerous biological materials, for in the last analysis, success in the realm of safety depends almost

entirely on the activities of the individual practitioners. But
I feel constrained to point out that the regulatory systems will
be a series of compromises, will surely be imperfect, and are as
certain as the sunrise to be breached every now and again--all of
which places them squarely within the framework of human endeavors
in general.

III. CONCLUSIONS

Aside from the practical question of safety, there is a matter
of cosmic philosophy that has set many a mind to churning. This
is the question of whether we have the right to interfere in a
fundamental way with the long-evolved barriers against genetic
exchange between species and, especially, between lower (prokaryotic) and higher (eukaryotic) organisms.

One's response to this question is a highly personal, and I
would add, emotional one. My own view is that whereas I am
violently opposed on moral grounds to tampering with the human
genome, I do not generalize that opposition to all organisms. So
far as I am concerned, the evolutionary barriers are not of
transcendant metaphysical significance but are simply a matter of
evolutionary practicality. It seems to me, safety aside, to have
little philosophical significance *on a cosmic scale* whether we
help a given organism to acquire a useful function or whether it
evolves that function on its own. The Darwinian laws will, ultimately, prevail in either case.

I would like to close with the comment that while the history
of our species is marked by noble achievement, it is also replete
with egregious folly. The human endeavor has always been and will
always be imperfect--how boring if it were not. Dealing with the
prospects and the consequences of scientific inquiry and technological advancement is, like all human endeavors, yet another
monument to our imperfections. Hopefully, we are learning as we

go and will not destroy ourselves in the process. And, as others have pointed out, the challenge of recombinant DNA and other great scientific advances, such as nuclear power and sophisticated electronics, is not only to the scientist to act responsibly, but also to the society which, ultimately, must make the decisions as to how knowledge will be used. Short of stopping all basic scientific inquiry, the society cannot abdicate that responsibility. Moreover, the society will not accept the challenge of solving a difficult problem unless it is forced to. Suggesting that we stop and wait for society to act positively in a vacuum or to evolve the appropriate political context within which rational use for the common good will be made of scientific knowledge is illusory. The political evolution will only proceed hand-in-hand with the scientific. The area of recombinant DNA research is, so far, a milestone in that evolutionary process.

It is a situation in which there are manifest and speculative dangers but in which the society will not (and, I feel should not) attempt to cut off the basic line of inquiry, but instead will be forced ultimately to deal with the various possibilities that are raised by it, such as manipulation of the human genome, and in the meantime, is and should be forced to act so as to ensure the safety of the species and the biosphere while the basic inquiry is being conducted.

REFERENCES

Burkitt, D. P. 1971. "Epidemiology of Cancer of the Colon and Rectum." *Cancer*, *28*, 3-13.

Chang, S. and Cohen, S. N. 1977. "*In Vivo* Site-Specific Genetic Recombination Promoted By *Eco*RI Restriction Endonuclease." *Proceedings of the National Academy of Sciences USA*, *74*, 4811-4815.

The Dangers of Unrestricted Research:
The Case of Recombinant DNA
A Response to Novick

Michael Ruse

Department of Philosophy
The University of Guelph
Ontario, Canada

Richard Novick's paper is absolutely fascinating. He etches
in vivid detail the dangers attendant upon Recombinant DNA
research, showing loopholes that exist in present so-called
"safeguards," and conjuring up scarifying scenarios to illustrate
the gruesome potential that such research has. Indeed, I would go
so far as to say that in some respects Novick's paper is too
successful for its own good; more precisely, I would suggest that
Novick apparently does so good a job of showing how lethal R-DNA
research might be, that the reader is altogether unprepared to
accept his about-turn almost at the end of the paper when he
claims suddenly that "I believe the overall threat [from R-DNA
research] is not great enough to justify interdiction, even if
that were possible" (Novick, 1978, p. 30). This about-turn is
even the less convincing because Novick has just said that
"Controls are appropriate but are *certain* to be imperfect" (his
italics); he immediately thereafter says "let us act so as to

minimize the risk [of R-DNA research], although it is surely
impossible to eliminate it"; and he then promptly ends his paper
with a final ghastly scenario, in which deadly dill pickles are
poised, about to wipe out the human race! One has visions of
some super-criminal buying up shares in McDonald's in order to
destroy the youth of North America in one fiendish master-stroke.

In short, I feel that if one just had Novick's paper to go
on, one would be a fool not to agree with someone like Robert
Sinsheimer that the dangers of unrestricted R-DNA research, even
with safeguards, are so great that one simply ought not permit it.
With friends like Novick, what need has R-DNA research of enemies?
(I suspect that at this point I am being a little unfair to Novick,
because presumably he was asked to contribute to the volume by
addressing the dangers of R-DNA research; hence the somewhat lop-
sided nature of his paper. However, I do not think I am being
unfair, because as his paper stands Novick has no right to draw
the conclusions that he does.)

So how best can one comment on a paper such as Novick's.
Obviously the answer will depend upon whether one agrees with his
conclusion or not. Can a case be made for the legitimacy of R-
DNA research, despite all the dangers? I take it that no one is
going to argue that a case can or should be made for totally
unrestricted research, in the sense of freely allowing anyone to
try anything; for instance, letting any mad scientists try to
create some new and absolutely lethal organism, like Novick's
Clostridium botulinum, able to grow in acidic environments. The
important and interesting question is whether a case can be made
for the legitimacy of R-DNA research, bound by the kinds of safeguards that scientists and governments are presently trying to
construct, but essentially unrestricted otherwise? That is to
say, R-DNA research which is allowed to go under its own momentum,
but done under careful containment conditions and where the
creation of deliberately hazardous organisms for their own sake
is barred.

My own feeling is that such a case for the permissibility of
R-DNA research can be made, and therefore by way of commentary on
Novick's paper--because I do not feel that he has made such a
case, although he wants one--I am going to sketch what I think are
the outlines of such a case. I realize that what I am doing may
seem, and is indeed, a little presumptuous. We are dealing with
scientific matters here and I am not a scientist. However, in
dealing with the problem of R-DNA research we are dealing with
something of concern to all of us--as taxpayers we are helping to
support it and as humans we are liable to be wiped out if it goes
wrong. Hence, for this reason I feel justified in contributing
to the debate. Also, as will be seen towards the end of this
paper, I believe the problem involves questions which are not
purely scientific, where a philosopher comes fully into his own.

1. R-DNA RESEARCH AS TECHNOLOGICALLY VALUABLE

The obvious place to begin making a case of R-DNA research is
with the positive goods that such research promises, in the form
of benefits for human beings, both the sick and the healthy members. (Although this technological argument is often presented as
being the only case which can be made for R-DNA research, as will
be seen later I do not think this is so.) Fortunately other
papers in this section have dealt at some length with the promised
benefits of R-DNA research, so it is not really that necessary for
me to say much more here. An obvious development which would
help all human beings would be a nitrogen-fixing bacterium, which
could grow in the roots of crops which at present need massive
doses of fertilizer. I do not think it is any great exaggeration
to say that probably the most urgent problem facing humans today
is the way in which we are running out of conventional forms of
fuel, particularly oil. But the production of artificial fertilizers, so necessary for modern farming, requires oil. Hence,

any way in which R-DNA research can reduce our dependency on oil,
cannot but be of immense value. (This is just looking at matters
from a global viewpoint. If we look at specifics, then the potential value of such a bacterium is even greater. First, poor
countries like India will not need to spend so much of their
foreign capital on oil imports. Second, the whole Western dependence on the oil-states will be reduced. For all the faults of
America, I have no wish to see its world-supremacy taken over by a
country as unenlightened as Saudi Arabia.)

The possibility of benefits of R-DNA research to society's
unhealthy members is perhaps even greater than that of the possibility of benefit to society as a whole. Clearly, for instance,
some newly created organism (or more likely, appropriately molded
older organism) capable of synthesizing insulin is going to have
inestimable benefit for the many diabetics in our midst. Similarly there is hope of medically valuable human proteins, like
clotting factors which might be used in the treatment of haemo-
philiacs. And of course this is not to mention the dread disease
hanging over us all, namely cancer. There is hope of help in
treatment, not to mention prevention.

It seems therefore that (without considering objections)
looking upon R-DNA research as a form of potential technology, a
very strong case indeed can be made for it. It offers hope of
unbelievably greater benefits for all members of the human race,
both healthy and sick. There may of course be disagreement between individual scientists about precisely how great a potential
R-DNA research may have; but it would be churlish to deny that it
is there.

2. PRELIMINARY OBJECTIONS TO
R-DNA RESEARCH

So much then for the positive case for R-DNA research considered as technology. If this were all that there is to be said,
then the matter would be closed at once. But of course it is not.
Major objections have been levelled against such research. The
important question for us is whether these objections are so great
that they outweigh the positive case to be made for R-DNA research.
The most obvious and gravest objection is that of the dangers to
human physical health inherent in R-DNA research. But before
turning to that there are two preliminary objections that ought to
be considered.

In the first place it might be suggested that all of the fancy
"benefits" that technology, including R-DNA research, is going to
bring are really not such a very good thing when looked at carefully. The claim might be made that we have grown soft from
industrialized civilization—sitting all day in overheated houses,
when we are not driving around at high speeds in gas-guzzling
cars—and hence perhaps the time has come, if only for our mental
health, when we ought to cry "stop" to such "progress."

I must confess that I have a great deal of sympathy for this
objection: Concorde strikes me as being a clear-cut case of technology run wild. Nevertheless, as a general position it grossly
overstates the case. In the last century, apart from a privi-
leged minority, most people lived in or close to abject poverty—
and all, rich and poor alike, were subject to the most dreadful
diseases: T.B., diphtheria, polio, and so on. Although science
and technology have not been the sole causes of the changes which
have so improved the life of so many, they have been major causes.
And I for one embrace the changes and hence the causes. This
really is "progress" in a good and genuine sense, and inasmuch as
R-DNA research can contribute to its continuation—and apparently
the prospect is that it can—in that sense I welcome it.

The second objection (to the argument that R-DNA research offers hope of good to humankind) is that, if our hope really is for the good of humankind, there are far better ways in which we might invest our time and money than on such research. There is so much human misery today for which the eradication depends, not on breakthroughs of new biological technology, but on the care and attention of men and women using what we know already.

Against this objection Maxine Singer has written in a *Science* editorial as follows:

> It was argued that all recombinant DNA research should be stopped and the nation should instead give priority to the distribution of existing methods of health care and sound nutrition to the citizens of the world. This argument is specious. Certainly there is a need to rectify inexcusable inequities in the distribution of food and health care. But recombinant DNA research is neither an alternative to such action nor a competitor for the necessary funds. Rather it provides for future opportunities to solve problems that are unsolved or that may yet arise. (Singer, 1977; p. 127)

One may feel perhaps that this rough dismissal of the objection is a little on the quick side (although obviously, editorials for *Science* have to be a little on the quick side!) Perhaps, as things stand, R-DNA research is not a competitor for the necessary funds. However, if our motivation truly is the general good of humankind, then perhaps we should be trying to get the funds for R-DNA research diverted to more worthy causes. After all, even government and university budgets are not fixed for all time, like the laws of the Medes and the Persians.

But nevertheless, Singer does make a point which if developed can effectively counter the objection that R-DNA funds should be diverted elsewhere. It would be foolish to put absolutely all of

our efforts into the immediate problems facing human beings. We
should indeed plan for the future, and here the case for R-DNA
research can be made. In fact, of course, to a certain extent the
future is now. Already we face a critical shortage of conventional
fuels. If, as has been suggested above, R-DNA research can go
some way towards alleviating this crisis--not just for us, but for
the Third World also--then it has as urgent a claim on our time,
resources, and energies as anything. Hence, I would suggest that
this objection also does not hold up.[1]

3. THE DANGERS OF R-DNA RESEARCH

The preliminary objections now considered, let us turn to the
major objection to R-DNA research: the dangers. Whatever the
benefits, can it justify the dangers? Before plunging right into
this problem, perhaps a tangential observation will be in order.
If indeed R-DNA research can lead to deadly results, can we afford
not to do it?

To date, those things with potential or evil which are fairly
readily available (*e.g.*, hand guns) have tended to be such as
could cause only a limited amount of damage (comparatively speaking), and those things that could cause great damage (*e.g.*, H-
bombs) have tended not to be readily available--although the
example of what India did with the reactor sold by Canada shows
that the last part of this claim is already a little dated, as
also does the case of the ship missing with twelve tons of uranium
on board. But, R-DNA research promises to change the whole picture. Even if it is an exaggeration to claim--as has been claimed
--that two men and a boy can do R-DNA research unaided, clearly it

---

[1] I consider this kind of objection in the related context of
genetic counseling in Ruse (1978).

and its results are more readily available than was any of the
technology and materials needed for making super-bombs. And, to
be honest, the thought that someone like Idi Amin might get his
hands on the deadly fruits of R-DNA research rather boggles the
imagination. The consolation that, if past experience is anything
to go on, the first people upon which he would use such fruits
would be his own people is no consolation.

So where then does this leave us, knowing full well that even
if the U.S. calls an immediate and total ban on DNA research, it
is going to continue to go on in many other places all over the
world? Can the U.S. afford *not* to do recombinant DNA research:
not so much as to come up with super biological weapons of its
own--it has enough weapons in its arsenal already--but to find out
what the potentialities for harm really are, and if they are real,
what if anything, can be done about them. A plausible case might
be made here for such work for the sake of national defense--
using the term "defense" here to mean just that, and not a euphe-
mism for another way of killing our fellow humans. I must confess
that, as a liberal, I feel a little uncomfortable about making
such an argument such as this. Somehow, it seems too reminiscent
of the 1950s and the Cold War. But, because such an argument
would come readily from the lips of John Foster Dulles, it does
not necessarily follow that it is unsound. In fact, I rather
suspect it may have some strength.

But this is all a bit tangential. Let us now grasp the major
nettle, namely the question of safety. As I have said earlier,
no one (certainly not I) wants to argue for totally unrestricted,
unbounded R-DNA research. The point is, whether even with sensible safeguards of the kinds which have been proposed, the positive
case for R-DNA research still holds up. Judging purely on the
basis of Novick's paper, I am far from sure that it does. R-DNA
research seems just too dangerous. But, despite Novick's
impressive scientific qualifications, is this the last word on the
matter? I am not convinced that it is, although I realize that in

saying this what we might have here is a classic case of a
philosophical fool rushing in where scientific angels fear to
tread. Certainly however it is the case that there are experts
who are prepared to argue that the dangers of R-DNA research have
been greatly exaggerated. And, having considered what people like
this have to say, I am inclined to agree to the extent of concluding that the dangers of R-DNA research do not outweigh the benefits. (As admitted, I am in the paradoxical position of disagree-
ing with Novick's arguments but agreeing with his conclusion, for
he too concludes that the dangers of R-DNA research do not outweigh the benefits. The point is that I do not think he has shown
this, although perhaps the topic he was assigned precluded such a
demonstration.)

What I intend to do therefore is to appeal to scientists who
do not think R-DNA research too dangerous, to counter those who
do. Of course, for all of their much-vaunted "objectivity" it is
usually the case that with respect to issues of great scientific
interest one can find well-qualified people who take very different positions.[2] Hence, on general grounds one might not think too
much of such appeals to authority. However, if one can find an
expert who is respected (and whose ideas are taken seriously) by
people on both sides of the divide, or who argues for one position
even though he has been taken as an authority by people on the
other side, then perhaps he deserves especial respect. In the
R-DNA research debate there is at least one such expert, namely
Professor Bernard D. Davis of the Harvard Medical School. Although Davis is not himself working on R-DNA research, and hence
does not have a personal axe to grind, he has argued recently that
the dangers of such research have been much over-estimated. And
yet, Davis as the author of a respected textbook on microbiology

---

[2]Think of the sociobiology controversy, for example! (See
Ruse, 1977).

has been quoted as an authority by opponents of all such research,
for instance by Erwin Chargaff in a somewhat hysterical letter to
Science which concluded that our generation "has been the first to
engage, under the leadership of the exact sciences, in a destructive colonial warfare against nature. The future will curse us for
it" (Chargaff, 1976; p. 938, reprinted in *Appendix, this volume*).

So, a *prima facie* case having been made for the integrity of
Davis's ideas, what does he have to say? The most important point,
one which certainly impressed me most strongly, is that despite
their great scientific achievements, when it comes to talking
about dangers most of the molecular biologists involved in the
R-DNA debate are talking outside of their field! When we are
concerned about the dangers of the spread of lethal newly created
forms of life, we are no longer in the field of molecular biology,
but in that of epidemiology. However:

> Instead of recognizing that the familiar features of the
> organisms offered grounds for reliable, general epidemi-
> ological predictions (as we shall see below), even before
> laboratory studies on the new strains provided more pre-
> cise information, the scientists involved proceeded as
> though nothing relevant was known and they were peering
> into a black box. The resulting preoccupation with
> unlimited scenarios contributed to the anxiety that they
> came to feel--and then transmitted to the public. To be
> sure, in order to establish the pathogenicity of any novel
> recombinants with precision, direct tests would be neces-
> sary. But a purely empirical approach could not be suffi-
> cient, for the next experiment might still inadvertently
> unleash an Andromeda strain. Eventually, then, it would
> be necessary to take into account also what is known about
> epidemics--and this information could have been helpful
> from the start. (Davis, 1977; p. 549)

We turn therefore to epidemiology, seeking help with respect
to our queries about the dangers of R-DNA research.

4. THE ARGUMENT FROM EPIDEMIOLOGY

In considering R-DNA research from this viewpoint, we can
locate three points of potential risk: first that R-DNA research
will produce a potentially dangerous organism; second, that such
an organism will infect a laboratory worker; and third that such
an organism will spread beyond the laboratory. Let us consider
these three points briefly in turn.

As far as the first point is concerned, one could quite
possibly deliberately make a dangerous organism; but this mere
possibility is not at issue here. Present guidelines exclude
deliberate attempts to make the possibility an actuality, and in
any case our discussion revolves around dangers despite precautions being taken. Of more concern to us is whether inadvertently
dangerous organisms might be produced, for instance through
"shotgun" experiments where random fragments of DNA (including
human DNA) are introduced into *Escherichia coli*. Particularly
troublesome would seem to be human tumor virus genes being incorporated into *E. coli*. *E. coli* live in the human bowel, and the
last thing we would seem to want are organisms which can live in
humans and which contain chunks of cancer genes.

However, for a number of reasons the fears raised at this
point may not be that justified. Probably tumor viruses are going
to be rare--if so, the chances of being incorporated are slight.
If they are common, then presumably a few more will not make that
much difference! Moreover, there is probably no such thing as a
cancer virus *per se*--the problem is that under certain circumstances certain genes give rise to cancers--these genes in other
circumstances are "normal." And finally, and most importantly,
analogously in nature shotgun experiments are probably going on

all the time anyway. DNA in the gut, released from dead cells,
will sometimes be picked up by the bacteria there. Hence, in the
human case, there will have been selection to protect us against
bacteria containing chunks of human DNA. In short, Davis argues
that with respect to the first point, the problem of dangerous
organisms, research within safeguards seems fairly safe.

Before going on to the second point, one should perhaps in
fairness to Novick interject what I think would be an objection
by him here, namely that "one can easily do a calculation to show
that only an infinitesimal fraction of all possible genetic combinations could have been tried throughout the course of evolution
thus far" (Novick, 1978; p. 11). Hence, we might indeed create
some new lethal organism. The answer is that of course we might,
but that the chances are very slim. This is because selection
will have given us protection not against the individual shotgunned
bacteria we have so far encountered, but against the whole class
of such bacteria (virtually by definition it has to do this because all of the bacteria will be unique). In other words,
although newly created or altered bacteria are unique, we have
protection against their kind. The situation is analogous to a
psychiatrist when she or he is presented with a new patient. She
or he will never have encountered someone with precisely that
past history or perhaps even symptoms, but her or his general
training will have prepared them to deal with people of that kind.

We come to the second point, possible illness in a laboratory
worker. It is at this point that the opponents of R-DNA research
start to get really frenzied, because they feel that the choice
of *E. coli* on which to do experiments is playing with fire.
Chargaff writes:

> I shall start with the cardinal folly, namely, the choice
> of *Escherichia coli* as the host. . . . If our time feels
> called upon to create new forms of living cells--forms
> that the world has presumably not seen since its onset--

> why choose a microbe that has cohabited, more or less
> happily, with us for a very long time indeed? (Chargaff,
> 1976; p. 938)

There are a number of points which can be made here to quiet
the fears of the critics, but two points will suffice. First, in
one very important sense *E. coli* is a very safe organism to work
with. Instead of floating through the air and then infecting
people--something which requires very careful precautions against
infection--it has to be swallowed. Swallowing is much less likely
to occur accidentally. Second the *E. coli* of the laboratory today
(quite apart from the fanciest new strains), is just not the *E.
coli* of the human gut. (It is known as *E. coli* K12). In its
fifty years in the laboratory, it has been selected right away
from its ancestors.

> Indeed, the present product has such radically altered
> surface properties that if it were isolated today it
> might well not be classified as *E. coli*. Moreover,
> this adaptation to laboratory media has occurred at
> the cost of deadaptation to the human gut: in recent
> tests with a large dose in man (much larger than could
> be expected from a laboratory accident) cells of this
> strain ceased to appear in the stools after a few days.
> Thus *E. coli* K12, outside the laboratory, is like a hot-
> house plant thrown out to compete with weeds. (Davis,
> 1977; p. 552)

Finally, we come to the third and perhaps most important point
of all. What are the risks of escape from the laboratory and of
an epidemic? Again however we should not overestimate the
dangers and raise fears. By adding more DNA to an *E. coli*, we
are not going to turn it into something other than it is, namely
an *E. coli*! (We are augmenting the DNA by about .1%.) Hence, if

the organism is to spread and cause disease, it has to spread and cause disease in the way that *E. coli* spreads and causes diseases, namely through the water supply and the like (*i.e.*, it will not suddenly become airborn like influenza). But, already we have good sanitation techniques designed to stop the spread of *E. coli* and similar organisms--chlorine in the water for instance. In other words, the chance of danger from a virulent form of *E. coli*, even if it escaped, is slight indeed.

So what conclusion can be drawn.

I conclude that illness in a laboratory worker caused by a recombinant organism would require compounding of three probabilities, all low: that a dangerous gene will inad- vertently be incorporated in the EK2 cells, that a large number of cells will be taken up, and that these cells will cause harm despite their short survival. Hence the danger appears to be vanishingly small, compared with the very real danger faced continually by medical microbiolo- gists working with known virulent pathogens--organisms well adapted to survive and produce disease. (Davis, 1977; p. 553)[3]

[3]EK2 strains of *E. coli* are those which have been developed because they are unable to survive and reproduce without arti- ficial laboratory conditions. Although Novick speaks depreci- atively of some of them, they do surely add a further safety factor.

5. THE VALUE OF R-DNA RESEARCH CONSIDERED AS TECHNOLOGY

I admit that the conclusion that R-DNA research is not really that dangerous is, on the evidence presented, the conclusion of one man. And indeed, by Davis' own admission, his assessment of low risk to R-DNA research is "not conventional." However, against this I would remind the reader that because Davis is taken as an authority by both sides, he deserves especial attention. Moreover, Davis makes a most important point: one which is certainly missing from Novick's paper. The real people to judge the dangers of R-DNA research are not necessarily the scientists doing the research, paradoxical as this may sound, but people who have experience of the spread of disease: epidemiologists. Moreover, they, unlike the molecular biologists, are not working in a vacuum of ignorance. They have had experience of disease and its spread. Even if one disagrees with Davis' conclusions, it is on his grounds that the battle ought to be fought, although one must add that even though he allows his conclusion of low risk is not conventional, he defends it as being not idiosyncratic.

> For despite public impressions to the contrary, the experts
> are *not* in sharp disagreement over this issue--if we define
> the experts not as just any scientists, but as those familiar
> with epidemiology and infectious disease. In my own contacts
> with colleagues in this field I have encountered none who
> consider *E. coli* recombinants a serious hazard, compared
> with known pathogens; . . . (Davis, 1977; p. 553)

Parenthetically, I must confess that to me, an outsider, Davis strikes something of a note of sanity. We all revel in the limelight and I suspect that some of the R-DNA researchers (and their opponents) are revelling a bit more than most. For years, biology has been the poor relation of physics, which has got the brightest

students, the largest research funds, and the most attention.
But now the focus of public consciousness has switched to
biology,[4] and quite naturally biologists are delighting in magni-
fying the glamour and danger of their work, spinning phantasmagoric
Ray Bradbury scenarios. As science fiction it makes good
reading, but if we are going to assess the real dangers of R-DNA
research--done under proper safeguards--let us come back to Earth
and turn to the great experience of scientists who have long been
working with dangerous organisms.

   My conclusion thus far therefore is that I see great potential
benefits from R-DNA research and that I cannot see that the
dangers outweigh these benefits. I argue consequently that R-
DNA research is an acceptable, not to say morally praiseworthy,
enterprise. I qualify this conclusion, as always, by saying that
reasonable safeguards should be applied and one would certainly
want to disbar deliberate attempts to create dangerous organisms
for their own sake. However, it would be wrong to end the discussion abruptly at this point, for we have as yet considered only
part of the pertinent argument. To this point deliberately I
have been considering R-DNA research as *technology*, that is to say
that I have been considering it as something which could possibly
bring great benefits to human beings. But this is in fact only
part of the argument. R-DNA research is also *science*: that is to
say it is part of the general inquiry of humans to understand
themselves and their environment. Because of this, I think we
have yet another powerful argument why R-DNA research ought to be
permitted. Let me expand on this.

[4]*Time* magazine cover stories, no less!

6. THE VALUE OF R-DNA RESEARCH CONSIDERED AS SCIENTIFIC ACTIVITY

Human beings are animals. Some of us may not like this fact very much, but it is true nevertheless. However, we are more than just animals. To be a human being is to be a thinking, inquiring being: one who explores the world, both within and without. Although Professor Sinsheimer has argued that we ought not play God, let us not forget that in an important sense man is little lower than the angels. To this fact we owe our greatest literature, painting, music, philosophy--and science. Hence, it follows that stilling creativity and the desire to explore, whether it be that of a painter or a biologist, is a dreadful thing and is in some very real sense a denial of humanity. Putting the matter bluntly: the unexamined life is not worth living.

Now, this very basic fact has a direct bearing on the R-DNA debate. Forget for a moment all about the technological side to R-DNA research and the great benefits that it is supposedly going to bring. Concentrate on R-DNA research as science, as the exploration for knowledge. As science it is, I believe, about as exciting as science has ever been, for through R-DNA research we are really stripping life down to its barest elements and understanding how living things work. I realize of course that even the R-DNA researchers themselves do not often speak of themselves as inquirers in the quest for pure knowledge--after all, they are dependent on substantial grants from public and private funds, and hence it is in their own self-interest to emphasize the technological payoffs of their work. But essentially, a major thing that drives R-DNA researchers is a quest for knowledge and the sheer excitement of being at the boundaries (together, no doubt, with the hope of Nobel Prizes and the like!). And what the R-DNA researchers are producing (as a whole) is first class science.

But if R-DNA research is indeed what I have just claimed it is--a manifestation of the human thirst for knowledge--then in

itself it is a good thing and ought to be cherished. Not because
it is of technological value, but because in doing it we are being
human in a proper and praiseworthy sense. This vital fact is
often overlooked in the R-DNA debate, if not pushed to the background, but let us not ignore it. As we look back through human
history we praise and value Archimedes, Copernicus, Galileo,
Newton, Darwin, because they, like Plato, Dante, Michelangelo,
Shakespeare, Tolstoy, expanded the human dimension. We should not
be too timid or cynical to extend this praise and value to R-DNA
research.

In arguing as I do, namely that science and its activity has
an intrinsic worth, and that this extends to R-DNA research, let
me not be misunderstood. I am not suggesting that any scientist
at any time or place has the right to do precisely what he or she
pleases. Creativity and intellectual exploration are not incom-
patible with responsibility--certainly the scientist has no right
deliberately to pursue a course of action that will probably cause
harm. For this reason, the argument I am making does not rule out
a call for safeguards. However, one must be careful not to mag-
nify unreasonably this call for responsibility. If one tries hard
enough, one can think up dreadful consequences for just about any
scientific activity, including R-DNA research. The fact that
such scenarios are possible, does not mean to say that they are in
any sense probable. As in all human activity, we must balance the
good against the bad. One might get killed in a plane crash, but
this does not mean that it is unreasonable to fly. Similarly,
for R-DNA research one must try to assess the possible dangers,
and if these do not really seem that likely--and I have suggested
that perhaps they do not--then I think that one must allow R-DNA
research because in itself it is a good thing.

Of course, no less than scientists, philosophers disagree,
and I doubt every philosopher would go all the way with me. One
that I suspect might disagree is Hans Jonas, who has written
extensively on scientific responsibility, and hence in order to

counter possible objections and to sharpen my own position, I
shall conclude this paper by looking very briefly at his ideas.

7. JONAS ON THE SUPPOSED DISTINCTION
BETWEEN SCIENCE AND TECHNOLOGY

Basically, I think Jonas would object to the way in which I
have drawn a distinction between pure science and technology,
following which I was able to argue that, whatever the technological aspects, a case could be made for the value of R-DNA research
considered as science. Although Jonas feels that at one point
one could make this distinction, perhaps back in Greek times, he
feels also that no longer is the distinction viable. The rise of
modern science has "entirely altered the traditional relation of
theory and practice, making them merge ever more intimately"
(Jonas, 1976; p. 15). Indeed, "not only have the boundaries
between theory and practice become blurred, but the two are now
fused in the very heart of science itself, so that the ancient
alibi of pure theory and with it the moral immunity it provided no
longer hold" (Jonas, 1976; p. 16). Particularly since the Industrial Revolution scientific advances have been spurred by and
evaluated in terms of their potential technological payoffs.

Hence, to Jonas the science-technology distinction is no
longer a really valid one. All science now is essentially a kind
of glorified technology. Moreover, argues Jonas, we have further
evidence that science has come to be a kind of technology because
of the nature of its aims: ". . . it has come to be that the
tasks of science are increasingly defined by extraneous interests
rather than its own internal logic or the free curiosity of the
investigator" (Jonas, 1976; p. 16). In fact, claims Jonas, not
only has science become a kind of technology considered from the
external viewpoint of what it is intended to do or serve, but
even within itself it is a kind of practical activity. When we

reflect on how science is done, "it is then borne in on us that doing science already includes physical action; that thinking and doing interpenetrate in the very procedures of inquiry, and thus the division of 'theory and practice' breaks down within theory itself" (Jonas, 1976; p. 16).

The conclusion that Jonas draws from this line of argument is that pure science no longer exists, and thus the scientific enterprise cannot be defended on grounds of an unsullied quest for truth. It must be evaluated and controlled as we evaluate and control all human activity, in terms of benefits and harms. Certainly, the consequences of science bring us into an area to do with morality, and even the very activity of science does now. The kind of argument that I have given, that science is to be valued because it is a free human inquiry into knowledge for its own sake, no longer makes sense. It is an anachronism.

Furthermore, argues Jonas, if one persists in drawing a distinction between the activity of science and its consequences, one cannot legitimately defend the activity of science as such on the grounds that even though its practice involves activity, such activity itself is morally innocent. Traditionally the case has been made that the activity of science does not involve the whole world, but only small scale experiments and these only on innanimate objects. But this is no longer true. Today, scientists experiment on a large scale, using the whole world as a laboratory.

> And as to experimentation on animate objects, which came
> with the younger biological sciences, no surrogate will
> do, no vicarious model, but the original itself must serve,
> and ethical neutrality ceases at the latest when it comes
> to human subjects. (Jonas, 1976; p. 17)

It must in fairness be noted that in the work to which I refer, Jonas does not address himself directly to the R-DNA research debate controversy. However, it is easy to see how his

ideas readily apply. He would say that R-DNA research is essentially like a technology, it must therefore be judged as such, and
cannot properly be defended in the way I have just been defending
it, namely as something morally praiseworthy in itself. "To the
public authority of [the general court of ethics and law] even
the vaunted freedom of inquiry must bow" (Jonas, 1976; p. 17).

8. REPLY TO JONAS

There are a number of points which can be made against an
argument such as that of Jonas, and I believe their sum effect is
to show that my arguments and conclusion still stand. As a preliminary though I would note that, even accepting Jonas' argument,
my position is that R-DNA research is permissible, because judged
as technology I believe the potential benefits outweigh the dangers. But this is only a preliminary because I want also to make
the case for R-DNA research on grounds that Jonas would deny.

In reply, first let us clear away Jonas' last point. I agree
fully that there are some scientific activities which are in themselves ethically unacceptable. Certain kinds of human cloning
experiments for example. However, R-DNA experiments (*i.e.*, those
within safeguards) do not fall within this category. For instance,
no one wants to make poisonous *E. coli* and then to feed it to
unsuspecting humans. R-DNA experiments are done on a small scale
and they are not done in such a way deliberately to threaten
human life or happiness. The mere fact that they involve living
organisms like *E. coli* rather than inanimate objects is quite
untroublesome. One has no more moral obligation to an *E. coli*
than one has to a pendulum. Hence, inasmuch as R-DNA research
itself involves doing things, as opposed to just thinking things,
no pertinent issues are raised.

Second, I would deny the way in which Jonas collapses science
into technology: at least I would certainly deny the kinds of

conclusions he seems to want to haul out of such a collapse. Of
course, it cannot be denied that much science has practical applications and consequent dangers, although it might be noted that
it is a little invidious to single out science on this score.
Much the same case can be made against philosophy. For example,
Karl Marx's work has had at least the practical consequences that
any scientific discovery has ever had. So on general grounds we
should be wary of arguing that science demands some special controls or is in some way deserving of peculiar restrictions or
censorship. Otherwise, in the name of consistency we may find
ourselves clamping down on any free inquiry in the name of the
potential dangers it might have. (On the philosophical front I
should think a paper like this would be amongst the first to go,
since I am obviously proposing a *carte blanche* for deadly
research!) There are very few human activities--scientific or
otherwise--which do not carry some threat of danger, and if we
are not to give up the vital human activity of inquiry, then some
risks must be taken. But as I have also said, I do not see that
in the R-DNA case the risks are so very high as to call for
special comment or total ban.

Yet, in a sense, this is all beside the point. I have argued
that R-DNA research is a good in its own right because it is a
prime instance of free human inquiry. The important question is
whether in pointing out that science has practical applications
Jonas has negated this point. I do not see that he has. Whatever
the practical implications of R-DNA research it is still a genuine
scientific activity producing genuine science. It is not merely a
sophisticated technological endeavour, and therefore only worthy
to be evaluated on those grounds. Jonas might deny this, for we
know that he argues that "the tasks of science are increasingly
defined by extraneous interests rather than its own internal logic
or the free curiosity of the investigator." In short, he seems to
think that in content, as well as in consequence, modern science

has become technology. But this is just not so; particularly not
in the case of R-DNA research.

To be perfectly honest, I am not quite sure what the "internal
logic" of science really is, but if we look most obviously at
history of science then what we see is an increasing effort to
bring the world beneath natural law, and moreover, although vital-
ists do not much like this fact, effort to connect everything up,
particularly in the sense that larger things are explained in
terms of smaller things. That is to say, one sees a kind of
reductive trend in science. In physics, macrophenomena are
explained in terms of microphenomena. Chemistry is explained in
terms of physics. Macrobiological phenomena like organic characteristics are explained in terms of smaller things like genes.
And of course today we have a major controversy as the social
sciences are being brought kicking and screaming into contact with
the biological sciences, the controversy over so-called "sociobiology" (See Ruse, 1977).

Now, whatever one might think of it, R-DNA research stands
about as firmly in this tradition as it is possible for something
to be. What we have is inquiry on the borderline between the
physical sciences and the biological sciences, as scientists are
finding out in the most basic way precisely what makes the world
work, seeing if in some manner the gap can be bridged between the
inorganic and the organic domains. Can organic macrophenomena,
like phenotypes, be explained in terms of inorganic microphenomena,
like molecules (see Ruse, 1973)? If this is not a case of science
driven by its own internal logic, I do not know what is! Moreover,
since the whole area of inquiry stands at such a key point on the
overall scientific reductive plan, it would be ridiculous to deny
the genuineness of the inquiry of the scientists concerned. This
is inquiry in the most fundamental and mainstream sense.

In short, what I argue is that R-DNA research does count as
scientific activity in a proper sense and is therefore in its own
right worth cherishing. Since its technological spin-offs are

also of value, and the dangers seem an acceptable risk, I argue
that R-DNA research (with proper safeguards) ought to be allowed
and encouraged.

REFERENCES

Chargaff, E. 1976. "On the Dangers of Genetic Meddling."
    Science, 192, 938-40. Reprinted, *this volume, Appendix.*
Davis, B. D. 1977. "The Recombinant DNA Scenarios: Andromeda
    Strain, Chimera, and Golem." *American Scientist, 65,* 547-55.
Jonas, H. 1976. "Freedom of Scientific Inquiry and the Public
    Interest." *Hastings Center Report, 6* (August), 15-17.
Novick, R. 1978. "The Dangers of Unrestricted Research: The
    Case of Recombinant DNA." This volume.
Ruse, M. 1973. *The Philosophy of Biology.* London: Hutchinson.
    _____. 1977. "Sociobiology: Sound Science or Muddled
    Metaphysics?" In F. Suppe and P. Asquith (eds.). *PSA 1976,*
    2, 48-73.
    _____. 1978. "Human Genetic Technology: the Darker Side."
    In *Problems in Biomedical Ethics.* Edited by J. E. Thomas.
    Toronto: Hackett.
Singer, M. 1977. "The Recombinant DNA Debate." *Science, 196,*
    127.

The Pathogenicity of *Escherichia Coli*

Samuel B. Formal

Department of Bacterial Diseases
Walter Reed Army Institute of Research
Washington, D.C.

Escherichia coli is an inhabitant of the intestinal tract of many animal species and of virtually all human beings. This group of microorganisms is important to this symposium because one strain, *E. coli* K-12 (or a derivative of it), is the leading candidate to be the host which will harbor the various recombinant DNA molecules constructed by molecular biologists. Concern has been voiced over the possibility that the introduction of foreign DNA into *E. coli* will transform this organism into one capable of causing disease of epidemic proportions in human beings. In view of this concern, I have been asked to discuss the potential for *E. coli* to produce disease with special reference to strain K-12.

The usual levels of *E. coli* found in normal human adults are approximately one million to ten million cells per gram of feces. These organisms constitute only a small fraction of the total microbial flora (less than 0.1 percent) of the stool; organisms which grow only in the absence of oxygen are far more prevalent. Studies have shown that three or four distinct strains of *E. coli*

reside together in the bowel; they remain and multiply for
periods of two to four months being replaced from time to time by
other *E. coli* strains. The factors which are responsible for this
colonization are not known. In addition to the resident *E. coli*
flora, transient strains contaminating our food and water appear,
but these do not persist and are isolated from the stool for only
short periods of time.

It is difficult to predict how a particular *E. coli* with
normal cell-wall components will behave when introduced into the
gastrointestinal tract of a given individual. Such an organism
may not be isolated from the stool or it could become a resident
strain. As an example, Table 1 gives the results of an experiment
(done in collaboration with Dr. R. B. Hornick's group at the University of Maryland School of Medicine) in which an *E. coli* strain
originally isolated from a healthy laboratory worker was fed at
two dose levels to adult volunteers.

TABLE I. *Duration of Shedding Following Ingestion of E. coli*
          Strain HS by Healthy Volunteers

Volunteer	Dose (cells)	Duration of Excretion (days)
1	1×10^{8}	12
2		0
3		16
4		60
5		45
6	1×10^{10}	105
7		21
8		7
9		14
10		75

It is evident that multiplication occurred and in some individuals
the organism was excreted for a long period of time. In contrast,
Dr. E. S. Anderson in Great Britain obtained different results

when he fed comparable doses of the common laboratory strain *E. coli* K-12 to volunteers in England. None of the individuals shed this particular strain for more than seven days. Strain K-12 is deficient in cell-wall components and either it or mutants derived from it will be used as a host for recombinant DNA experiments.

While most strains of *E. coli* are considered to be non-pathogenic, others do cause disease. *E. coli* is the most common cause of urinary tract infections and is isolated with increasing frequency from the bloodstream of hospitalized patients with underlying illnesses. The special attributes which *E. coli* must possess to cause urinary tract infections or bacteremia are only now being studied. However, some general statements can be made. The source of *E. coli* which causes these diseases is the intestinal tract—frequently that of the patient himself. Thus, survival in the bowel is a necessary prerequisite for causing disease in a significant number of individuals. As already mentioned, strain K-12 has only limited ability to remain in the gut and present evidence indicates that *E. coli* K-12 mutant strains can be isolated which fail to survive at all. In addition, other properties are required for pathogenicity. A common attribute of all disease-producing *E. coli* is the presence of an intact lipopolysaccharide (LPS) component. Strain K-12 has a defective LPS, making it more sensitive to the defense mechanisms of the host. Other cell surface components which allow the organisms to attach to the wall of the urinary tract or to survive in the blood stream are also required to cause infection. Strain K-12 also lacks these surface components.

Some strains of *E. coli* cause diarrheal disease and there is specific information available concerning the pathogenesis of this process. To induce disease, the organism either must be able to multiply in the small intestine and elaborate an enterotoxin or must be able to penetrate the intestinal epithelium and multiply in the tissues. When these diarrheal disease mechanisms were defined, attempts were to confer pathogenicity on originally

avirulent *E. coli* strains. Dr. H. Williams Smith in England
transferred both the ability to elaborate K-88 antigen (required
for the organism to reside in the small intestine of piglets) and
the ability to elaborate enterotoxin to certain avirulent strains
of *E. coli*. He showed that these laboratory-constructed organisms caused diarrhea in piglets. However, when these same two
virulence factors were incorporated into *E. coli* K-12, this strain
failed to multiply and remained nonpathogenic. Clearly, additional attributes are required to convert *E. coli* K-12 into a
toxigenic pathogen. Our group at Walter Reed has been attempting
to prepare safe oral vaccines against bacillary dysentery. We
have transferred the ability to synthesize cell wall components
of virulent *Shigella flexneri* 2a to *E. coli* K-12. Not only did
this hybrid strain fail to cause disease, but when fed to volunteers, (again in collaboration with Dr. Hornick), it was shed
(Table 2) in the stool to no greater extent than was the wild-
type K-12 strain which was administered by Dr. Anderson to volunteers in England.

*TABLE II. Duration of Shedding Following Ingestion of E. coli
K-12 - Shigella flexneri Hybrid Strain*

Volunteer	Dose (cells)	Duration of Excretion (days)
1	1×10^8	0
2		0
3		3
4		4
5		3
6	1×10^{10}	0
7		5
8		4
9		4

Thus, at present two different groups have not been able to transfer virulence to *E. coli* strain K-12 by introducing genetic material from known pathogens. It is obvious that virulence is determined by a multitude of genes all functioning in concert. It therefore seems highly unlikely to me that the random inheritance of DNA from widely diverse species will transform *E. coli* K-12 into a pathogen which will cause epidemic disease in the United States. I state this with a considerable confidence on the basis of past experience. A large number of scientists in the United States have worked over the years with virtually every known pathogen. Laboratory infections have occurred. A relatively small number have resulted in the deaths of laboratory workers. Yet I am aware of no instance where disease has spread to the surrounding community.

While this is reassuring, we should not feel a false sense of security. Until we know more of the potential dangers of recombinant DNA research, the experiments which are carried out must be carefully considered. For instance, it has been suggested that a cellulose-producing *E. coli* strain be constructed which, following its colonization of human beings, would allow them to utilize cellulose as a nutrient. If we put aside any inherent risks in doing such an experiment, I submit that it should not be conducted for the simple reason that it is based on two false assumptions. Firstly--as mentioned previously--purposeful colonization of individuals by *E. coli* is unpredictable, and if it does occur, does not last for more than a few months. Secondly, in human beings, the bulk of *E. coli* resides in the large intestine, and this is the place where nutrients from cellulose fibers presumably would be made available for absorption. But the human large intestine is not an organ for the absorption of food and little if any nutrient enters the body from this region.

Certainly, we should continue to have some concern over what effect recombinant DNA molecules might have on the pathogenicity of *E. coli* K-12 or its derivatives. However, we must remember

that this foreign DNA is conserved in its weakened *E. coli* K-12
host in the form of a plasmid. Under certain circumstances plasmids can leave their host cells and infect other bacteria. Therefore it seems more important to me to ensure that plasmids carrying recombinant DNA cannot escape their K-12 hosts, infect other
hardier bacteria, and thus have a greater opportunity to survive.
This aspect of the problem is under active study by several groups.
Its satisfactory resolution should allow recombinant DNA research
to proceed at a more rapid pace than it has in recent years.

Ethics

Ethical Prerequisites for Examining Biological Research: The Case of Recombinant DNA

Daniel Callahan

The Hastings Center
Institute of Society, Ethics, and the Life Sciences
Hastings-on-Hudson, New York

Despite the cumulative experience of the past two or three decades of coping with rapid technological change, the debate on recombinant DNA indicates how little progress has actually been made. As a society, we do not know what to do with technological advances which promise both benefit and harm, nor do we know how to carry on an adequate debate at either the scientific or the political level. It is as if, despite considerable political experience with nuclear energy, environmental problems, and a host of other technological innovations, we have learned practically nothing at all. The case of recombinant DNA is particularly difficult because we do not have at hand ready moral principles to deal with issues of this kind, a body of social and ethical experience to resolve such issues, and very little in the way of a social or ethical methodology for grappling with the issues. Hence, it appears, we must devise almost *de novo* a methodology and decision-making procedure.

It would be foolish to expect that one could achieve anything
approximating a precise decision-making procedure. The issues are
too confused to allow for a simple deductive approach, where
specific decisions can easily be derived from generally acceptable
premises. The best one can hope for would be the establishment of
procedures which would, at best, help delineate and clarify the
issues, and push decision-making in one broad direction rather
than another. I want to attempt here to lay out a framework for
such a procedure.

At the political level, it is now possible to discern the
development of a few moral and political premises which are
increasingly commanding general assent. Few, that is, are likely
to condemn these premises in public places, e.g., before Congressional committees.

Here are the premises:

(a) scientists have a moral responsibility for the consequences of their work, even in basic research;

(b) the public has a right to intervene in scientific research,
including basic research, when it is paying the bills for that
research, and/or when that research could conceivably affect the
welfare of the public, either for good or for ill (but especially
the latter);

(c) in line with that last-mentioned right, the public also
has a correlative right to use the normal mechanisms of public
policy to express its wishes and interests--up to and including
the right to pass restrictive legislation on research.

I will not here attempt to establish the validity of those
principles, though I believe they are valid. I am only asserting
that they now seem to command wide assent, and that they underlie
recent efforts on the part of Congress, states and municipal
governments to inquire into, and possibly to regulate, recombinant
DNA research.

Even if those three premises can be taken as established, the central ethical question remains: by what ethical criteria should the public judge biological research, and particularly recombinant DNA research? It is simply not enough to say that the public has such rights. It is necessary also to inquire what constitutes a responsible use of those rights. Obviously, if the right of intervention and control is pressed too far, considerable damage could be done to the future of scientific research. At the same time, if those rights are wholly neglected, the public will then have no means of playing a role in the development of that scientific research which could directly affect its own welfare. Neither extreme is acceptable.

Much of the debate on recombinant DNA has been cast in terms of the language of comparing and weighing benefits and harms likely to result from the work; the language of "trade-offs" is commonly invoked. I believe this approach, taken by itself, promises to get us practically nowhere. First, with the possible exception of pure knowledge resulting from recombinant DNA research, both the practical benefits and concrete harms are wholly speculative. No numerical probabilities can be attached to those benefits or harms, and therefore the usual cost-benefit methods cannot be applied. Second, even the speculative benefits and harms are asymmetrical: the worst scenarios envisage ruination of the biological and evolutionary order, while the best scenarios envisage only benefits, not human salvation. Hence, the possible harms are, in principle, of an order and magnitude quite incommensurable with the possible benefits. Third, even if it were possible to more completely and empirically compare the benefits and risks, there is no evidence of any social consensus whatever on what would constitute an acceptable risk, or what would constitute a benefit worthy of gambling with various degrees of risk.

A calculus of risks and benefits can only be employed where there exists substantial social agreement on the use of such a calculus, and an agreed-upon set of norms for balancing risks and

benefits. If this is, however, true in general with any technological issue, it may nonetheless be less true if this problem
can be brought within a more familiar arena of moral discourse.
Let me attempt to place the recombinant DNA debate within that
more familiar arena.

One common method of trying to cope with a large socio-moral
issue is to ask, of any proposed course of conduct (public or
private), how it would serve to foster or hinder those moral and
social values already accepted in society.

A. HUMAN FREEDOM AND SELF-DETERMINATION

Does the research promise to enhance, or at least not to harm,
the value of human freedom and self-determination? Will it enhance
the freedom of all, or only of some? At best, what could it contribute to freedom? At worst, what could it take away from human
freedom? What are the possible consequences for the self-
determination of future generations? Is it possible for us to do
things now which would be irreversible by future generations?

In order to pursue this line of questioning, it would be
necessary of course to make a distinction between freedom to do
something, positive freedom, and freedom from certain intrusions
or incursions, negative freedom. In general, one would look at
every envisioned benefit and risk of recombinant DNA research, and
ask whether those advances would be freedom-enhancing, or freedom-
reducing.

B. JUSTICE

Do the fruits of the research, under favorable conditions,
increase or decrease the possibility of a just society? Would the
benefits of the research benefit all, or would they only benefit

some: Would any deleterious results of the research fall upon all
equally, or only upon some? Would the public have the possibility
of controlling the allocation of both the benefits and harms of
such research?

C. SECURITY AND SURVIVAL

Does the research promise to enhance human security (e.g.,
security against illness, famine, etc.) or could it harm security?
Is there any possibility that the research could threaten human
survival itself, or the survival of a significant number of
people, or the survival of a particular social group? Would the
chances of survival be enhanced by a favorable outcome of the
research?

D. THE PURSUIT OF HAPPINESS

Would a favorable outcome of the research enable people to
live happier, more satisfactory lives (defining those phrases any
way one wishes)? Might it jeopardize that possibility?

The values of freedom, justice, security and survival, and
the pursuit of happiness by no means constitute a definitive list
of values, but it is sufficient to suggest the kind of ethical
methodology which might be used. Yet there is a real difficulty
with using such a methodology by itself. It would still involve,
on a case-by-case basis, a balancing of costs and benefits, and
the problem would be compounded when one began making comparisons
among the values. Security and survival might be endangered, but
individual freedom enhanced. What would we do then? Nonetheless
as I will argue below, it can be a useful method if complemented

by some other considerations. Those I will call "moral policy
considerations."

I do not believe it possible to deal effectively with the
ethical problem of recombinant DNA research unless one has established a moral policy toward research. What do I mean by a "moral
policy?" I do not mean the establishment of precise decision-
making procedures, or the establishment of some rigid hierarchy of
values, but rather this: the establishment of a general attitude
and set of biases which will in general incline a person, or a
social body, in one direction or another, and which will coherent-
ly color the way in which evidence is interpreted and costs and
benefits weighed.

Here are three "moral policy" options:

A. INTERVENTION OR NONINTERVENTION INTO NATURE

One can take the general position that mankind should not
intervene into nature at all unless it is absolutely necessary
for the sake of the preservation of human nature to do so.
Alternatively, one could take the general position that mankind
should feel perfectly free to intervene into nature as long as it
suits our purposes and meets other moral tests. Nature, that is,
can be construed as either sacred or in some sense wise or both,
and intervention approached with great wariness. For instance,
it might be contended under such a policy that long-established
evolutionary patterns in nature should not be tampered with.
Alternatively, it might be argued that nature is neither sacred
nor wise, and that intervention is perfectly appropriate if it
promises to produce no practical harms.

B. RISK-TAKING VERSUS CAUTION

Here again one encounters a basic choice: one can decide that,
all other things being equal, it is part of the glory of human
nature (and specifically of science) that it is willing to take
risks, and willing to gamble in order to understand life better
or to improve life. Scientific progress is both possible and
good, and progress is always to be preferred over the status quo,
even if it is necessary to take some risks in the name of progress. Alternatively, one could take the position that caution
and care is the best policy, that, all things being equal, it is
best to adopt a stance of wariness, and to take only slow,
measured steps toward change and innovation.

C. DOING GOOD VERSUS AVOIDING HARM

Again, a fundamental kind of moral choice: ought the goal of
the moral life, taken at least in the social and community sense,
be that of seeking always to do good--to make life better, to cure
illness, do away with misery, make more people happy; or alternatively, is our moral obligation limited to the avoidance of
doing harm to others?

The main point about these three broad moral policy options
is that, in each case, the choice of one option rather than
another will make a considerable difference in our inclinations,
in the way we interpret problems, in the way we understand evidence, and in the way we are inclined to handle specific moral
dilemmas. And I have not, in this case, quite chosen these policy
options at random. For it has been argued, against recombinant
DNA research, that science should not meddle with something so
delicately wrought as evolutionary patterns, that is, we should

not cross well-established species lines. It has also been
argued, in favor of the research, that even if there may be risk,
we should be willing to gamble for the sake of attaining great
benefits, that progress is only gained by taking risks. In short,
the choice of one of the competing moral policies rather than
another can make a decisive difference, if not independently of
other considerations, certainly in a way which can decisively
influence what we make of those other considerations.

There is something else to be noticed as well. I have
deliberately cast the policy options in dichotomous form: intervention versus nonintervention, risk-taking versus caution, and
doing good versus avoiding harm. If we combine the various
options together, the overall drift of a policy will be even more
determinative. A combination of a general policy of intervention, combined with a policy of risk-taking, combined with a
policy of doing good, will almost certainly incline one to favor
moving forward on recombinant DNA research. If this would be true
even in cases of biological research where it might be possible to
assign some specific probabilities to risk and benefits, it would
be even more true in those cases--as in recombinant DNA research--
where one can compare only speculative benefits and risks. Conversely, if one's policy choices combine resistance to intervention, plus a policy of caution, plus a policy of avoiding possible
harm--then, one would almost surely be inclined toward stopping
the research or going ahead very slowly and carefully.

I have so far argued that a simple ethic of comparing benefits
and risks will not work in the case of recombinant DNA research.
I have also suggested that a more traditional route testing these
supposed benefits and harms against a list of specific values
might make the moral issues more manageable. I did, however,
point out that such a method brought one back again to some of the
same problems encountered in trying to use a simple calculus of
comparing costs and benefits. I want now to argue that only the
development of a broader moral policy could allow us in any

effective way to make an efficacious use of the specific values
we wish to foster or protect. We cannot, that is, simply ask
about the research whether freedom, justice, and so on, are
enhanced or hindered without asking such questions in the context
of a broader moral policy, which would allow us to attach some
weight and value to the specific questions about the specific
values. Indeed, I think it would be impossible in any meaningful
sense to make much sense of a list of values without a moral
policy or policies in hand.

Let me give some examples. *Freedom:* It is not enough simply
to ask, particularly in a speculative situation, whether the
research will promote or hinder freedom. We must also ask--as a
broader policy question--just how much freedom (or justice, or
security) we are willing to gamble in order to enhance freedom.
Do we want to hold on to the freedom we have, or gamble and seek
more? *Justice:* It is not enough to ask whether advancement of a
technology will promote or jeopardize justice. We must also ask
the broader policy question of whether we are willing, say, to
risk potentially hazardous interventions into nature in order to
promote justice. For it is quite open for someone to say,
plausibly enough, that if we intervene into nature in a dangerous way, then we may altogether remove or obviate the possibility
of justice (a major plague, for instance, could not be expected
to promote the value of justice). I will not multiply examples.
Their point should be evident: our moral policies will and should
influence our assessment of whether the technology will foster or
hinder the realization of values we think important. Those
values, in turn, cannot themselves be used or interpreted apart
from the context of a moral policy or policies. This is only to
say that we cannot ask of the research whether it advances or
retards the development of particular values in a vacuum. The
point of a moral policy is to provide a context for asking the
more specific questions.

I have by no means presented here a decision-making procedure. In the present circumstances, which includes most basic biological research in general and recombinant DNA research in particular, there exists no decision-making procedure. One should, moreover, be suspicious of any simple solution of the problem of decision-making. In the end, we will simply have to make a choice, and there will be--as there almost always is when a final decision must be made--a gap between our rational considerations and the taking of that final step which is called action. What I am proposing is a way of acting rationally and prudently; it is a sorting procedure, in the end, as distinguished from a procedure which tells us how to act in specific circumstances.

But it is a procedure which suggests that the major part of our thinking must be done at the policy level. We must, in the best sense of the term, try to determine where our biases should lie, and thus in what general direction we should be inclined. This requires that we come to some broad (for they could be nothing less) conclusions concerning, among other things, whether we think that interventions into nature are, *on the whole*, beneficial or dangerous; whether we think that the taking of risks has, *on the whole*, proved itself to be worthwhile; and whether we believe that a morality which seeks to do good will, *on the whole*, promote the human good better than the one which simply tries to avoid doing harm. These seem to me the real questions, at least in those circumstances when we are admittedly--as with recombinant DNA research--taking steps into an unknown future.

It is only natural at this point to inquire just how we are to reach and justify broad standards concerning the problems of intervention, gambling, and doing good. Are we not simply forced into even more general and abstract questions? Of course we are, and even worse, there exist still fewer procedures to allow us to ask how we should choose among the various dichotomies I have mentioned. While it would be a mistake to think that the making of such choices is inherently a-rational, it is probably necessary

to concede that we are probably here dealing with a choice among
fundamental stances toward human life itself. People can and do
read the history of scientific advance and technological application very differently--at one extreme are those who think that
life has really not been improved at all, over against others who
think that life is now far better than in earlier generations
because of scientific advance. It is exceedingly hard to adjudi-
cate those arguments, and it seems clear to me that the determining factors become broad categories of metaphysical thinking
which cannot be fully established to everyone's complete satisfaction. The question of whether one should approach life, and
nature, and scientific research, in the spirit of gambling and
risk-taking is not one, I would suggest, which can simply be
decided by adding up whatever the evidence is.

I believe it possible to interpret the political debate over
recombinant DNA research in terms of the categories of moral
policy I have suggested here. If one takes the guidelines
established by the National Institute of Health as one key indi-
cator, one can see in those guidelines a middle-ground policy,
and compromise between and among the dichotomies I have suggested.
In essence, those guidelines say that some types of recombinant
DNA research are at present too risky, that one should not
intervene into nature in certain cases, but that, on the whole,
it is acceptable to intervene into nature with recombinant DNA
research. Moreover, the guidelines imply that it is worth taking
some risk, so long as those risks are approached slowly, prudently, and in terms of constant outside review and monitoring.
Finally, the guidelines imply that it is, on the whole, better
that research go forward in order that the possibility of gaining
the speculative benefits might be achieved. The value of continued scientific research is thus affirmed, and in the process it
is equally affirmed that it is better to adopt a policy of
attempting to do good rather than a policy of simply avoiding
evil, in which case a permanent moratorium might have seemed more

appropriate. Put another way, one can interpret the guidelines
as refusing to make black and white dichotomous choices. On the
contrary, the guidelines in effect exclude the most radical possi-
bilities--simply going ahead without any controls, or simply
stopping research altogether--and look for a way of taking the
dichotomies into account, but not allowing polar positions to
dominate. In an American political context, one might interpret
this as a classic example of brokerage, compromise politics--where
an attempt is made to take account of polar positions, but, in
the end, to achieve a policy which reflects those polarities without taking them too seriously. Both those who think the research
is too hazardous to be accepted at all, and those who think that
no controls whatever are appropriate, will be unhappy with middle-
ground, compromise guidelines. Brokerage politics rarely satis-
fies everyone--its purpose is to find a *modus vivendi* between
opposing factions, a policy option which will satisfy as many
people as possible, and one which will, if not satisfy those at
the extremes, at least reduce their complaints to mutterings.

In the case of recombinant DNA research, a major distinction
concerning political morality is necessary. It does not seem to
me that a public decision is necessary to morally support a nonintervention into nature, or a policy of caution, much less a
policy of doing no harm. On the contrary, a policy to move for-
ward--to attempt to do good, and to gamble--does require a positive public assent. It is not possible to make even a *prima facie*
case that there is a moral obligation on the part of any scientist,
or the scientific community in general, to pursue recombinant DNA
research. It may be desirable to pursue such research, but I do
not see how a case can be made that it is morally obligatory to do
so. The moral obligation could only exist if there was very
strong evidence that a failure to pursue the research would
jeopardize human existence itself. No one has tried to make that
extreme an argument. Hence, recombinant DNA research is simply
one of a number of optional lines of scientific, social or other

forms of research which may or may not be pursued, depending upon
what society counts as beneficial to itself. There might be a
loss to human progress if the research is not pursued, but it is
difficult to see how there could be any claim made that a failure
to pursue the research would be in itself immoral.

However, if a choice is made to move ahead with the research,
there is an obligation on the part of the scientific community to
gain the assent of the public. For in that case, a set of moral
policies will have been chosen which, on the face of it, have
implications for the public rather than simply for the individual
researcher--it is *our* world which is the object of intervention,
our lives which may be gambled with, and *our* benefits which are
alleged. Clearly, then, it is *our* decision to make, not that of
the individual scientist, or the scientific community taken
collectively. To paraphrase Garrett Hardin in another context, if
there is to be mutual benefit it must be mutual benefit mutually
decided upon, and if there is to be mutual risk, that mutual risk
must be mutually agreed to.

Comments on Callahan

Roy Curtiss III

Department of Microbiology
University of Alabama Medical Center
Birmingham, Alabama

Dr. Callahan enumerated three moral premises at the beginning
of his paper which he accepts as established. I by-and-large
agree with these, although there may be a need to qualify the
first; that is, that scientists have a moral responsibility for
the consequences of their work, even in basic research. Scientists do have an obligation to conduct their research in a manner
so as to minimize harm to themselves and, if at all possible, to
preclude harm to others. However, we need to distinguish between
research versus the uses of knowledge that stem from that
research. It is thus unclear that the scientists who develop the
recombinant DNA technology and acquire information from its use
should be blamed if society decides to use that technology and
knowledge in a manner which in the eyes of scientists may be
morally irresponsible and over which the scientists have little
control. If the scientific community is, however, to be held
responsible for the uses of knowledge which they acquire, it
could be argued that scientists would have to retain proprietary

149

rights over that knowledge in order to intervene to preclude
society from making irresponsible uses of it. Society could not,
however, accept this option, since by their financial support of
the research, the knowledge becomes public property. It therefore follows that society must accept moral responsibility for the
ultimate uses of knowledge and that scientists will have to accept
moral responsibility to provide factual information and advice
that will assist society in the wise and beneficial use of that
knowledge.

In another part of Dr. Callahan's paper he discusses various
moral and social values that could be used in evaluating the benefits versus harms of recombinant DNA research. I would agree that
freedom and self-determination, justice, security and survival and
the pursuit of happiness are important human values. However, he
has omitted one value that scientists hold very dear; namely, the
value of knowledge. I should also note that the acquisition of
knowledge is not a conjectural benefit of recombinant DNA research.
It is a well-established fact, as is revealed by perusal through
the recent scientific literature. I would therefore hope that as
society evaluates the benefits and risks of recombinant DNA
research and established moral and ethical principles to guide
their decisions, they would give careful consideration to the
value of knowledge.

Dr. Callahan has offered three general moral policies as a
means to establish general attitudes and biases to guide decision-
making procedures on and ethical considerations of recombinant
DNA activities. The first of these was "intervention or nonintervention into nature"; and in considering this, we must again
distinguish between research to acquire knowledge and the subsequent use of that knowledge. Geneticists and molecular biologists
have always tried to intervene into nature whether to develop
better breeds of chickens or to unlock the secrets of life.
Recombinant DNA technology certainly facilitates such an intervention approach. I do not believe, however, that the use of this

new technology as a research tool constitutes a moral issue of any
consequence to society so long as harm to others is minimized or
precluded. I should hasten to add, however, that some scientists
may choose to consider that using a research technique that intervenes into nature is contrary to their own moral philosophy and
thus decide not to use such approaches. This, however, should be
an individual decision and not one dictated by society. The use
of this knowledge to genetically engineer microorganisms, plants,
animals, and even humans is, however, intervention into nature
that should be subject to moral consideration by society.

The remaining general moral policies considered by Dr. Callahan
are "risk-taking versus caution" and "doing good versus avoiding
harm." I will deal with these together by describing some history,
much of it personal. In 1972, I left Oak Ridge National Laboratory
for the University of Alabama Medical Center where I could learn
more about bacterial infectious diseases and commence to study the
genetic and biochemical bases of bacterial pathogenicity. I intuitively believed then that plasmids conferring drug resistance
would soon be rampant in bacterial pathogens and that a new
approach would be needed to contend with bacterial diseases.
Recombinant DNA technology provided an approach to study these
microbes, many of which had no classical system of genetic analysis. Paul Berg and a group of ten other scientists including
Herbert Boyer were at the same time considering the potential risks
of recombinant DNA research using strains of *Escherichia coli* and
various plasmid and phage cloning vectors. In July, 1974, their
letter to *Science* and *Nature* was published in which they asked for
a voluntary cessation of certain experiments, caution on others
and the convening of an international meeting to consider the
problem. I read this letter and soon became very apprehensive. I
ultimately wrote a lengthy open letter in which I enumerated many
"half facts" concerning the origin, prevalence, and attributes of
plasmids and about the pathogenicity and sexual promiscuity of *E.
coli*. I say "half facts" since I ignored frequencies of occurrence

and made no distinction between various virulent and avirulent
strains of *E. coli*. Since I initially believed the potential
biohazards to be real, I suggested a cessation of essentially all
recombinant DNA experiments until the biohazards could be assessed
and the means to contend with them established. I also began to
think of how to make the research safer and came upon the idea to
employ enfeebled strains of *E. coli* K-12. These ideas were
refined during further discussions with Drs. Novick, Falkow, Cohen
and Clowes and presented at the International Meeting on Recombinant DNA held at the Asilomar Conference Center in Pacific Grove,
California in February 1975. At Asilomar, we heard many talks.
E. S. Anderson told us that pathogenicity of a microbe depends on
three attributes: ability to colonize an ecological niche,
ability to display virulence to overcome host defense and communi-
cability. Data were presented on the inability of *E. coli* K-12 to
colonize humans and I talked about *E. coli*'s sexual promiscuity.
I noted, however, that transfer of non-conjugative plasmid cloning
vectors was an improbable event and showed a slide listing seven
or eight natural barriers to conjugational plasmid transfer that
are operative in nature. Some scientists intuitively reasoned
that the introduction of foreign DNA into *E. coli* K-12 was
unlikely to change all of its defects and convert it into a pathogen and reasoned from my data that transmission of recombinant
DNA to other microbes was likely to be very rare. They thus began
to believe that the risks were minimal. Others, like Robert
Sinsheimer, Richard Novick and myself were not at all convinced
that that was so. I then intensified work on the design and construction of safer *E. coli* K-12 hosts for recombinant DNA
research and all nine of my postdocs and students joined in the
effort. As a member of the NIH Recombinant DNA Molecule Program
Advisory Committee, I began an intermittent but sometimes intense
education of myself on all areas of science that seemed relevant
in assessing risks of recombinant DNA research. Upon completion
of our construction of the "self-destructing" *E. coli* host strain

χ1776 in January 1976, we commenced a rather exhaustive set of
tests that are still in progress. These tests on χ1776, which
was certified as an EK2 host in December 1976, and on various
"non-self-destructing" EK1 strains caused me to begin to think in
terms of probabilities of survival and probabilities of transmission of recombinant DNA to other microorganisms. We showed
that certain of the mutations in EK1 hosts diminish survival in
the rat intestine and that *E. coli* strains from patients, sewage
and polluted rivers are essentially always smooth--that is, they
produce a normal lipopolysaccharide in their outer membrane--
rather than rough like *E. coli* K-12, and usually do not have
nutritional requirements like most EK1 and EK2 hosts. Since a
normal lipopolysaccharide is requisite for pathogenicity and since
χ1776 could not survive passage through the intestinal tract, I
came to believe that there was no risk with any permissible experiment in using the physical and biological containment as specified
in the NIH Guidelines for Recombinant DNA Research. I always
qualified this statement, which was put forth at various meetings
this past winter and spring, with my concern that the weak link
was the likelihood for human error. More recently I reexamined
the entire issue of potential biohazards in using *E. coli* host-
vector systems. I first examined the reasons why *E. coli* K-12
cannot colonize the bowel and why it has been so far impossible
to endow it with this ability or with virulence by using conventional genetic procedures to introduce genes from known pathogens
able to colonize the gut and able to cause disease. I then
examined whether the introduction of a piece of foreign DNA using
recombinant DNA technologies could allow *E. coli* K-12, to colonize
the bowel or some other ecological niche, to display virulence, to
cause physiological harm and to become more communicable. I concluded that even if one of these lost abilities could be regained,
it would be insufficient to endow *E. coli* K-12 with pathogenicity.
I next used our extensive data on transmission of plasmid DNA to
conclude that transfer of foreign DNA to other bacteria, which is

more likely to occur in the gut than outside the body, is so
improbable that it will seldom occur. Besides, there is evidence
to strongly suggest that foreign DNA confers a selective disadvantage on the perpetuation of a vector and on a host containing it
and it is known that a vector, in the absence of selection for it,
is gradually lost from bacterial host cells. These observations
imply that such recombinant DNA, even if transferred to another
microbe, would not be likely to survive. I next examined various
scenarios for introduction of potentially harmful genes with the
help of experts in gastroenterology, pharmacology, endocrinology,
infectious diseases, etc. as well as all conceivable types of
human mistakes that could be made in a cloning experiment. These
analyses led to the belief that risks in all these instances were
at worst minimal but more often nonexistent. I am therefore convinced that the use of *E. coli* K-12 host-vectors for permissible
experiments poses no threat whatsoever to humans or other organisms with the exception that an investigator or lab worker who is
very careless may sometimes suffer harm. The basis for my conclusion is contained in a lengthy letter to Donald Fredrickson,
Director of NIH, and I have some copies available for those who
may wish to read it. I should hasten to add, however, that
although I am now convinced that harm cannot be a consequence of
introducing foreign DNA into *E. coli* K-12, I do not expect other
scientists to be so convinced until our data and analyses are published and they have an opportunity to scrutinize them.

Although I have not evaluated the potential safety associated
with using host-vector systems other than *E. coli* K-12, I have
spent two and a half years of my scientific life in taking the
cautious approach to provide safer systems for cloning and to
establish to my complete satisfaction, at least, that no harm will
come from this research unless it is a conscious decision of
society to use the knowledge gained from recombinant DNA research
for purposes over which the scientific community has no control.
I therefore wonder, if there are no risks and no harms to avoid,

whether we need moral policies for and ethical consideration of the research. Rather, these would seem to be needed to guide us in the wise and just use of the substantial knowledge likely to be gleaned from these studies.

In closing, a word or two about legislative regulation of recombinant DNA activities. Federal legislation is needed, if only to make the NIH Guidelines or some slight modification thereof pertain to all parties, public or private. I am very concerned, however, that legislation will be adopted as a political expediency and based on fear and ignorance. Repressive legislation which supplants reliance on and trust of scientists and institutional biohazard committees to ensure safety, with requirements for inspections, licensing, specific liability, etc. will hardly be a just reward to a scientific community that by its past actions should have earned that trust. May society therefore take *its* turn to act responsibly in ensuring that Congress not punish the non-guilty and legalize antiintellectualism.

The Limitations of Broad Moral Policies

John Richards

Department of Philosophy
University of Georgia
Athens, Georgia

In his lead paper for this section, Daniel Callahan argues
that the adoption of a broad moral policy is necessary for the
resolution of, among other issues, the recombinant DNA controversy.
I reject this proposal. Broad moral policies are too general to
be of use in specific controversies. In particular cases, *ceteris
paribus* clauses render a broad policy ineffective. Specific
issues need to be considered on their merits, and are rarely, if
ever, resolved by an appeal to broad policies. On the other hand,
a narrow focus on technical issues crowds out moral considerations,
and reduces decision making to a blind comparison of statistics--
numerology in the worst sense.

Controversial issues are marked generally both by the failure
of broad moral policies to provide direction, and by the limitations of technical data gathering. Both approaches fail to
address the actual difficulties, one is too broad, the other too
narrow. As a mean between the excesses of broad moral policies,

and the tunnel vision of data gathering, a more moderate infusion
of moral considerations into a technical controversy has the best
chance of producing a usable policy.

There already is a broad moral policy operating in science:
Proceed with caution, but proceed nevertheless. Its limitations
are quite apparent in connection with recombinant DNA research;
this is what I propose to show here in my reply to Callahan. In
section one I argue that, although everyone in the debate accepts
this policy, it has not proven effective in resolving issues.
Since each controversy is unique, an appropriate resolution must
focus on special features of that particular controversy. In
section two I argue that there are four special characteristics of
recombinant DNA research which must be taken into account in the
determination of policy regarding the research. In section three
I show that the NIH Guidelines do not adequately take into account
these special features of recombinant DNA research. Moreover, the
Guidelines were not derived from a rational policy. They were
developed to replace the Asilomar guidelines, under the pressure
of bringing the moratorium to an end and allowing research to
resume. They were promulgated in the void left by the inadequacy
of the broad moral policy of science. In section four I argue
that the original rationale for imposing the moratorium did reflect
the specific characteristics of recombinant DNA research. It was
a move to gain time. And finally, in section five, I conclude that
the most appropriate course is still to keep the pace (rate of
progress?) slow. This will provide time to evaluate and understand properly the nature of the dangers, and time to evaluate and
understand properly the special problems of the research.

I. BROAD MORAL POLICIES

Callahan argues that it is not possible to deal effectively
with the ethical problem of recombinant DNA research unless one

has established a moral policy toward research (p.140 *this volume*).
He then sets forth three options which help in determining an
individual's broad moral policy. His paper aims at fostering the
development of some broad policy, rather than defending particular
policy options, or a particular broad moral policy.

This approach overlooks the extremely powerful broad moral
policy already operating in the scientific community, and shared
by the public at large: *Proceed with caution, but proceed nevertheless.* This broad moral policy has proved inadequate to the task
at hand and, moreover, has limited the debate to a consideration
of scientific and technical issues.

The pervasivensss of this broad moral policy is evident when
we consider Callahan's three policy options in terms of the actual
course of the debate over recombinant DNA research.

A. *Intervention or non-intervention into nature*

The option determining for a moral policy is whether or not we
should intervene in nature, and, if so, under what conditions.
But it has generally been accepted by all sides in the debate that
we should avoid an irreversible intervention in nature. The
pertinent question is the extent to which we are intervening. It
has not been at all resolved whether, and in which particular
cases, we might intervene in nature. Those who reject Sinsheimer's
arguments against transgressing natural barriers do so on
scientific grounds, not because they believe we ought to intervene in nature. It is argued instead that in nature itself this
barrier is often crossed. Therefore recombinant DNA research
does not involve a radical intervention, but repeats a commonplace
occurrence. The participants in the debate have not been arguing
the moral question: whether we ought to intervene in nature, or
the philosophical question: what is meant by "intervention in
nature?" There seems to be general agreement on these points.

Instead, the debate has revolved around the technical question:
are we in fact radically intervening in nature?

B. *Risk-taking versus caution*

The moral policy option is the extent to which we are willing
to take risks, to gamble, in order to improve life and to foster
scientific progress. The major argument in defense of continuing
research is that (with a few possible exceptions which are banned
anyway) we are not taking any risks.   There is nothing dangerous,
and therefore we need have nothing to fear. Robin Holliday's
(1977) argument, which calculates the possibility of disaster at
10^{-14}, is aimed at dispelling our fears, rather than encouraging
risks. The controversy has not been over the moral questions:
whether and to what extent should we take risks, or even what
would be an acceptable risk (e.g., is 10^{-14} an acceptable risk?)?
There is no suggestion of a disagreement in this area. Instead
the actual controversy concerns the technical question: Are we in
fact taking any risks?

C. *Doing good versus avoiding harm*

The option for moral policy is whether we should actively seek
to do good and make people happy, or whether we are obligated to
avoid doing harm. The scientific community, by its very nature,
has accepted a "do good" posture. And it is generally agreed that
in the absence of harm, it is a goal of science and technology to
make life better and to attempt to cure illness, and so on. Ray
Valentine's arguments (*this volume*) are not aimed at convincing us
that we ought to do good. Rather in describing the possibility of
transferring the nitrogen fixation property, he is pointing out
some of the good which might be done. Here again, the participants

in the debate have not raised the moral question: Should we
actively seek to do good?   Or, perhaps, can we accept certain
levels of harm, in order to produce greater good?   These issues
have not been raised, I suggest, because there is a broad consensus.   Instead we have controversy over the scientific questions:
How real is the possibility that we can do good?   In what areas
can we expect fruitful action?

These three moral policy options have not been the points of
contention in the controversy.   Strikingly, the debate over
recombinant DNA has been marked by almost no intrusion of questions
of morality.   Rather, the arguments have centered strictly on our
capacity to predict and understand the results of our actions.
The participants in the debate all seem to have accepted that, if
the opponents' scientific arguments regarding the dangers *were*
correct, then this *would* be sufficient reason to proceed or
desist accordingly.   Similarly, there has been almost no question
of the value of the potential benefits.   There appears to be
general agreement that a demonstration of serious negative consequences associated with recombinants, would be sufficient grounds
for postponing applications.

Moral considerations, in the form of alternative broad moral
policies have been, and are, extraneous to the argument.   The
debate involves different evaluations of the evidence provided by
the findings of science, or more accurately the lack of such evidence.   Under the pressure to resolve the debate these evaluations
have come to the fore, and moral questions have been eclipsed.

I therefore reject Callahan's approach.   In the recombinant
DNA controversy there already is a widely accepted broad moral
policy, and it is inadequate.   Broad moral policies work, para-
doxically, when there is no controversy, and when there is a
generally accepted theoretical framework.   While the current theory
may not be true in any absolute sense, it does provide a basis
for understanding.   This may not be the best possible basis, but
it is a framework which allows for reasonable predictions.

Successful predictions are one of the trademarks of an accepted
theory. However, when we enter an arena where there is no
explanatory framework, any conceivable outcome is equally possible,
as are quite a few inconceivable outcomes. And this is precisely
the issue. In the absence of a theoretical framework we have no
guide to suggest what is conceivable or reasonable. In the
absence of such a guide broad moral policies are useless. This
may not be the place to argue this more general point, but, in
principle, broad moral policies are inadequate. Controversy
arises when there is no accepted method for handling an issue.
The appearance of controversy marks the breakdown of a broad
moral policy. Thus, almost by definition, broad moral policies
are rarely, if ever, of service in resolving conflicts.

II. RECOMBINANT DNA RESEARCH

I now isolate four characteristics of this research which must
be taken into account in the determination of policy: Ignorance,
Irreversibility, Global Implications, and Activity. Because these
are crucial, much of the debate is already framed in terms of
these characteristics. I am merely letting them out of the box
(Pandora's ?). The analysis will clarify the structure of the
debate, and perhaps, provide grounds for a more adequate resolution.

A. *Ignorance*

There is a fundamental lack of information regarding the
research. The present theoretical framework in genetics is not
able to provide explanations or make reliable predictions concerning the outcome of this research. The research itself, the
methodology, extends beyond the accepted theoretical framework.

This eliminates the predictive capacity of our theory, and therefore, it eliminates the possibility of estimating reliable costs
and benefits. Broad moral frameworks do not operate in the
absence of these estimates. The most significant aspect of the
research is that we are, at best, hazarding guesses.

This point is not mitigated if we are now prepared to provide
a theoretical framework. The research was initiated in a state
of ignorance. It was conducted in the absence of a theoretical
framework which could provide acceptable explanations or reliable
predictions.

B. *Irreversibility*

The results of the research are irreversible. In one sense,
everything we do is irreversible. We are doomed to endure consequences of decisions and actions of the past. But the irreversibility of recombinant DNA research extends beyond the asymmetric
nature of time. As Sinsheimer points out, if we introduce a
viable recombinant organism in to the biosphere, there is no way
for us to retrieve it. There will be no opportunity for us to
alter this action. The organism enters into its own evolutionary
pattern, and we then have lost control over the future course of
events. This is not the case with most other scientific research.
If things go awry with recombinant DNA research, we will have
initiated a chain of events with unknown and potentially powerful
consequences, and we will have no means by which we can alter,
correct, or in any way counter what we have done.

C. *Global Implications*

The research has global implications. There is no sure way of
containing a mishap. Once there is an accidental release of a

viable recombinant organism, whether it is deemed dangerous or not,
the likelihood is that its spread will be global. This too presents a sharp contrast with most other scientific research. A
nuclear power plant catastrophe may slightly increase levels of
radiation around the world, but the large danger is to the surrounding community. In contrast there is no way of restricting a
recombinant organism to a continent, not to mention a one hundred
mile radius.

D. *Activity*

As Hans Jonas argues, we are caught in a dilemma: in order to
gain the requisite information to allay our fears, we must engage
in the research. We are driven into the research in order to make
intelligent decisions concerning the safety of the research. We
overcome our ignorance through a series of actions which pre-
suppose their own harmlessness. In Jonas' words, "the very deed
eventually to be decided on in the light of knowledge is already
committed in the night of ignorance in obtaining that knowledge"
(1977, p. 18). In order to dissolve our ignorance we must act in
that ignorance.

These four features distinguish recombinant DNA research from
other areas of science. A broad moral policy which applies to
the normal operations of science, and to non-controversial areas
is not capable of assisting in the resolution of the recombinant
DNA debate. Moreover, by focusing on these features we avoid
questioning the entire scientific enterprise. Considering this
issue by itself makes it possible to propose limiting certain
kinds of research without questioning whether we are breaching
the sanctity of some broad moral policy insuring absolute freedom
of research and inquiry.

There is *prima facie* evidence that the debate has in fact
focused on the special features of recombinant DNA research. The

scientific community has itself provided much of the impetus for the controversy, and this community does not often, if ever, raise these kinds of questions. Controversy has arisen within the scientific sector precisely because the familiar broad moral policy was not an adequate guide in situations where: First, we are fundamentally unable to make predictions concerning research. Second, we risk an irreversible chain of events which, third, has global implications. And, fourth, our ignorance, in turn is only alleviated if we conduct the research. We must develop a policy designed to handle these special features and not intended for a generalized moral policy for scientific inquiry.

III. THE NIH GUIDELINES--CAN IMPLIES OUGHT

Most participants in the debate have acknowledged the above four characteristics of recombinant DNA research, usually in some sort of initial statement concerning the state of our ignorance and the tentativeness of the proposals and assumptions. Nevertheless, these considerations have not had much effect on the various sets of guidelines proposed in the past few years. These have all been offered as an end to debate rather than as a product of debate. Debate has been foreclosed through an appeal to the Technological Imperative: *Can Implies Ought*. If we can do something, then we ought to do it. This locates the burden of proof on those who urge caution. In the absence of demonstrated hazards, progress, in and of itself, is a sufficient justification to continue research. This is particularly obvious when we examine the developments which led to the formation of the NIH Guidelines.

From the very beginning, the Guidelines were only a part of the task of the committee. In the 1974 Berg committee report, the director of the NIH was asked to establish an advisory committee charged with:

(i) overseeing an experimental program to evaluate the potential biological and ecological hazards of the above type of
recombinant DNA molecules, (ii) developing procedures which
will minimize the spread of such molecules within human and
other populations, and (iii) divising guidelines to be followed
by investigators working with potentially hazardous recombinant
DNA molecules. (Berg, *et al.*, 1974, p. 303)

The report from the Asilomar conference concluded that, little is
known about the survival of laboratory strains of bacteria, about
the effects of recombinants on this survival, about potential
infectivity of recombinant molecules, about transferability, and
so on. They recommended further research:

There are many questions in this area, the answers to which
are essential for our assessment of the biohazards of experiments with recombinant DNA molecules. It will be necessary
to ensure that this work will be planned and carried out.
(Berg, *et al.*, 1975, p. 994)

And finally the official charge to the NIH committee was in three
parts, one of which was to recommend research on the hazards:

The goal of the Committee is to investigate the current state
of knowledge and technology regarding DNA recombinants, their
survival in nature, and transferability to other organisms; to
recommend programs of research to assess the possibilities of
spread of specific DNA recombinants and the possible hazards
to public health and the environment; and to recommend guidelines on the basis of the research results. (*Federal Register,
39* (November 6, 1974), 39306).

But this research was never carried out. Instead our assessment of the dangers remains a matter of conjecture. Our *fear* of
the unknown lurks in the background as a weapon to force stricter
containment. Our *wish* for medical or agricultural miracles which

seem within our grasp has been utilized as a prod to leniency.
But there has been no attempt to deal with the state of our knowledge in a direct way. And in the absence of this research, there
is no cohesive rationale for the Guidelines.

The Guidelines have allegedly been constructed along a cost-
benefit model, and, by and large, the debate was focused on this
aspect. As Callahan so correctly points out, this approach
promises to get us practically nowhere. To the best of our
ability we project, or estimate, costs (hazards) and benefits, but
the invocation of the technological imperative has placed the
burden of proof on the hazards. And this leads to gross inconsistencies in the Guidelines.

On the one hand, perhaps the greatest benefit would be a cure
for cancer, but experiments with even moderate risk (and hence
more potent) oncogenic viruses are prohibited. On the other hand,
the greatest risk is the release into the biosphere of a viable,
aggressive, radically new organism which will colonize a niche
already taken, irrevocably alter the biosphere, and possibly upset
some delicate balance. But, as pointed out by Sinsheimer, experiments which can radically alter *E. coli* K12 by inserting DNA from
any invertebrate are being conducted in what is called "moderate,"
that is P2, containment. P2 is hardly containment at all, just
good laboratory procedures.

The NIH Guidelines have not been constructed along a rational
principle. They have not even been constructed as a compromise
between costs and benefits; instead, as I have suggested, they
have been constructed as a compromise between fears and wishes.
No one denies the promise in the recombination techniques, and no
one denies that we may radically alter the biosphere. But we
have been unable to move beyond these rudimentary remarks except
by appealing to a maxim that we ought to do what we can do. And
this maxim implies that we should proceed.

Donald Fredrickson, in his prolegomena to the Guidelines
remarks:

There are only limited data available concerning the expression of DNA from higher forms of life (eukaryotes) in *E. coli*
(or any other prokaryote). Therefore the containment pre-
scriptions for experiments inserting eukaryotic DNA into prokaryotes are based on risks having quite uncertain probabilities. (NIH Documents 1976, p. 8)

Nevertheless, these experiments are permitted. This is
largely because the Guidelines are concerned mainly with the
possibility of *known* infectious or dangerous substances expressing themselves. This assumes that the alteration of an environment of a gene has little or no effect on what is expressed by a
particular gene. This is an unsubstantiated wish, adopted in
order to allow research to proceed.

On the preceding page, Fredrickson explains the prohibition on
deliberate release of organisms containing recombinant DNA molecules: "With the present limited state of knowledge, it seems
highly unlikely that there will be in the near future, any recombinant organism that is universally accepted as being beneficial
to introduce into the environment" (NIH Documents 1976, p. 7).
But if intentional release of a recombinant organism is so controversial as to be prohibited, it seems that we ought to take
greater precautions to prevent the accidental release of *any* recombinant organisms. Yet (see above) work is permitted on recom-
nant organisms in only a moderate risk, EK1-P2 experiment. Certainly our intention not to release it into the environment has
very little to do with the possibility itself.

These risks are considered acceptable because the organisms
being manipulated are not related to any known direct threat to
humans. The guidelines are constructed around a fundamentally
human perspective. The danger to human beings must be demonstrated prior to proscription. This was pointed out early on by
Roy Curtiss. He argued:

I think we altogether place too much emphasis on the damage
that these recombinants could do to humans. We don't happen
to be an indispensable species in the biosphere. Plants and
other microbes and things are, and we depend on them. I
think we have to broaden our thinking of what organisms might
be at risk. Although we like to protect ourselves, I think
we have to consider the others. (NIH Documents 1976, p. 237)

The anthropocentric attitude is very apparent in Robin
Holliday's recent article, "Should genetic engineers be contained?"
The subtitle is informative, "The probabilities of the events
in a hypothetical scenario of a man-made epidemic resulting from
recombinant DNA research shows disaster to be closer to science
fiction than reality." Even without elaborating the long history
of the conversion of science fiction to science fact, there are
important reasons to doubt the rhetorical nature of the title
question. Conveniently, the scenario Holliday depicts is precisely the scenario that the writers of the NIH Guidelines had in
mind: A human viral epidemic results from an accident in the
laboratory. Accordingly, Holliday is able to cautiously estimate
the chances of this occurring at 10^{-14}. He concludes: ". . .
even if the product of the probabilities is underestimated by
several orders of magnitude, the chances of an accident from
studies in heterogenetics [his word for recombinant DNA research]
remains vanishingly small" (1977, p. 401). Holliday's argument is
based on a detailed and impressive account of the possibility of
a single kind of accident (actually a few similar ones). But
Holliday concludes that the chances of *any* accident in recombinant
DNA research is vanishingly small.

This is an unwarranted conclusion. The traditional problem
with our ability to project the impact of new technologies has
been limited by the failure of our imagination. We have just not
been able to think of the thing that would go wrong. In Holliday's
case this is compounded by the assumption that we need only attend

to obvious immediate dangers to human beings. This deliberately
ignores our interdependence with other species, and is a direct
consequence of locating the burden of proof on those who suggest
hazards.

We are struck with just one more symptom of our inability to
quantify many of the variables of which we are aware. This
inability has led to a rather perverse and arcane system of protection. Why are the standards adequate at P2, in fact so precisely adequate that any raising of P1 experiments to P2 would be
"overly stringent" (NIH Documents 1976, p. 6)? Or, conversely,
why are the precise P2 level standards "necessary" (ibid.)? In
short, what would a reasonable estimate of the hazards look like,
and what possible rationale could provide consistency? I could
never figure out why the Guidelines allow P1 + EK3 = P2 + EK2,
especially when we do not know what EK3 is. Is it merely because
3 + 1 = 2 + 2? This line of reasoning allows P0 + EK4, or even
EK5 and P-1, or P5 and EK-1. Perhaps more than anything else
this demonstrates the absurdity of trying to adequately assess--
quantify--our ignorance.

The NIH Guidelines had to be formulated, but the prevailing
broad moral policy in science was insufficient to justify a course
of action. In the absence of a cohesive rationale, and under the
pressure for a solution, debate was foreclosed by appealing to an
aphorism: Can implies ought--and the result is the Guidelines
as we know them.

IV. RESEARCH POLICY--*FESTINA LENTE*

The challenge to the Guidelines for recombinant DNA research
is a challenge to the abrupt foreclosure of debate signaled by
the invocation of the Technological Imperative. This challenge
was prompted by the four specific characteristics of recombinant
DNA research outlined in part II. These characteristics

interjected a new element into the scientific milieu: fear of our own minimally understood capabilities. It is this fear which produced the moratorium.

The moratorium itself was a foreclosure of sorts. Research was ended, temporarily, to allow for debate. This foreclosure was brought about through an appeal to a different maxim: *Festina lente*--Make haste slowly (Suetonius, A.D. 160).

In the past, scientists, acting as individuals, may have exercised individual discretion and refrained from pursuing certain research. For example, Einstein, in a letter to Roosevelt, suggested the possibility of the atomic bomb, but refused to work on it himself. But, at no time in the history of science have scientists themselves been so awed by their ignorance that, as a group, it was felt that it would be better to declare a moratorium on research.

This is in direct contrast with more recent reaction to technological innovations.

> Influential and thoughtful segments of the public have begun to ask whether technological progress is always for the best-- whether we can afford to wait until unforeseen side-effects of technological applications reach crisis proportions before we seek means to alleviate them on a piecemeal basis. (Brooks & Bowers, 1977, p. 229)

The continual appearance of unforeseeable side-effects is a constant reminder of the limitations of human knowledge. The tremendous potential in recombinant DNA technology has led us to extend this reminder to science itself. The threat of unforeseen side-effects seems very real when we think about the specific characteristics of the research. The simple fact of the matter is that *our ability to project the impact of new technologies is directly limited by the failure of our imagination*. In retrospect, it is always easy to explain what went awry. The difficult part is to anticipate the problem.

Senator Edward Kennedy, in his introductory remarks to *Kennedy Hearings 1976*, argued that the societal impact of unforeseeable side-effects is the major issue in recombinant DNA research:

As long as the development of technology outstrips man's capacity to understand the implications of that technology, the chances for serious errors in judgment are increased and the possibility for serious societal disruption is very real. (Kennedy Hearings 1976, p. 1)

These considerations brought forth the moratorium. But it is important to understand that the moratorium was not proposed as a product or as the end of a debate. It was proposed in order to create time for debate. It was never considered as a final step, but rather as a pause, as an alteration of tempo. This pause was intended to allow our capacity to understand the implications of our technology to catch up to the development of that technology. It was never intended as a challenge to that technology.

Contrast this foreclosure to permit debate, with a challenge to the one genuine broad moral policy of science. The latter, as a challenge, creates only controversy. Under the pressure of controversy we become anti-science, anti-progress and anti-technology, or we become incautious and the glory of science beckons us to unrestrained investigations. Progress alone eliminates the evaluation procedure entirely. Caution alone eliminates science. On the one side, the most innocent question raises the (inapplicable) charges of Lysenkoism. On the other hand, every advance is seen as an intrusion into the natural scheme of things, and, therefore, dangerous and unwarranted. This extreme polarization has served only to obscure the central issues, and has worked to prevent any reasonable approach to the problems.

In many respects the debate has ended. It has been foreclosed for one side by the technological imperative, for the other by an appeal to make haste slowly. The disagreement is in the sense of time underlying the moral policy. This is a matter of attitude, more than policy. But, if it is framed in terms of policy, we shall have an irresoluble conflict. Under the pressure of promulgating policies, we produce inconsistencies.

In the summary of the Asilomar conference, under "Types of Experiments," there is what would seem to be overwhelming reason to delay experimentation: "Accurate estimates of the risks associated with different types of experiments are difficult to obtain because of our ignorance of the probability that he anticipated dangers will manifest themselves" (NIH Documents 1976, p. 66). But this is followed immediately by, "Nevertheless . . .", and the force of "nevertheless" is to completely discount the warning.

V. CONCLUSION

Broad moral policies are beset by all the problems associated with attempts to consider the general case. Real issues, however, are always special cases, and our decisions are always context dependent. General commitments and beliefs are inevitably overridden in controversial cases. A broad moral policy will insure, above all, the adoption and defense of consistent positions. My rejection of the assumption of a broad moral policy is tied to my rejection of the appearance of consistency as the ultimate test of a moral outlook. In practice the drive for consistency according to any pre-specified set of criteria always ends in absurd, but consistent, positions.

In recombinant DNA research, where (a) the dangers are unpredictable, (b) irreversible, and (c) potentially global, and (d) where all indications point to the research itself as being the only path to illuminate these dangers, the single most important action we can take is to slow down. There is no need for haste. By waiting or slowing down, we only postpone benefits. I admit, if we can talk in these terms, there is the question of the cancer

patients who would not be cured in the ten years that we might be
postponing a solution for cancer. But, I find it difficult to
justify that this effort needs to be expended now. Or, in
Callahan's words, I do not see that a scientist is obligated to
carry on this research immediately, since five or ten years of
related research may shed more light on the real dangers. Consider in this connection, Curtiss's comments to the effect that
he personally postponed utilizing the recombinant DNA technology
in order to work instead weakening *E. coli (this volume).*

Waiting, as a specific policy, has important side-effects.
First, it permits the development of research directed solely
towards understanding the nature and degree of the potential
dangers. This was, after all, part of the charge to the NIH committee. Second, it permits the excitement of a new discovery, the
thrill of the chase, to be replaced by a calmer consideration of
the alternatives. Third, it provides the time for experts to
see the issues as others, laymen, are viewing them. This is paradoxical, but it raises an important issue. The expert is clearly
the one who knows a field best, but there is a myopia built into
the training process for any specialty. From the viewpoint of
one's own field it is good to intervene, take (always moderate)
risks, and (of course) do good. We assume a much more moderate
posture when examining the issues in non-related fields when we
are the laymen. We are always slightly suspect of other fields,
and alternative approaches. We always hesitate to place our
lives and our futures in the hands of others. (Those other guys
may be irresponsible. They will blow us all up!)

But my final recommendation is not for the complete cessation
of recombinant research. It is a recommendation that the pace of
research be slowed. Specifically, this can be accomplished
through establishing a limited number of regional centers for the
research. These can be carefully monitored, with significant
space alloted to research aimed at evaluating the hazards. Funding itself can support research directed towards safety and

evaluation, rather than practical applications. For the latter,
even if successful, will still be controversial and consequently,
not practical after all.

There is no urgency. Moreover, the sense of urgency itself is
dangerous. And it is intensified by shifting the debate to the
level of alternative broad moral policies. These fail to directly
reflect the nuances of the particular issue and force us to make
sweeping decisions: continue or stop. The debate is not resolved
by appealing to the noble scientific search for truth, for this
too requires time for more careful consideration. Even if the
search is worthwhile because of its yield in terms of the quality
of life, time will allow the broader perspective needed to judge
the potential effects of this research on the quality of life. Or,
perhaps the search is worthwhile as an end in itself, or as a path
to truth. This search then takes its place as part of a long
tradition. But this tradition is not inviolate, and here too,
time will allow us to assume a broader perspective. In Nietsche's
words:

> In some remote corner of the universe, poured out and glitter-
> ing in innumerable solar systems, there once was a star on
> which clever animals invented knowledge. That was the
> haughtiest and most mendacious minute of "world history"--yet
> only a minute. After nature had drawn a few breaths the star
> grew cold, and the clever animals had to die.
>
> One might invent such a fable and still not have illus-
> trated sufficiently how wretched, how shadowy and flighty,
> how aimless and arbitrary, the human intellect appears in
> nature. (*On truth and lie in an extra moral sense*, 1873)

ACKNOWLEDGMENTS

I would like to thank Roy Cebik, Bernard Dauenhauer, Nicholas
T. Fotion, Susan G. Hadden, John Messer, and Tony Nemetz for their
comments on earlier versions of this paper.

REFERENCES

Berg, Paul, *et al.* 1974. (Letter) "Potential Biohazards of
   Recombinant DNA Molecules." *Science, 185* (July), 303.
   Reprinted in *this volume: Appendix.*
Berg, Paul, *et al.* 1975. "Asilomar Conference on Recombinant
   DNA Molecules." *Science, 188* (June), 991-994. Reprinted in
   this volume: Appendix.
Brooks, Harvey, and Bowers, Raymond. 1977. "Technology: Pro-
   cesses of Assessment and Choice." In Teich, Albert H. (ed.),
   Technology and Man's Future, 2nd ed. New York: St. Martin's
   Press, 1977: 229-242.
Holliday, Robin. 1977. "Should Genetic Engineers be Contained?"
   New Scientist, 73 (February), 399-401.
Jonas, Hans. 1977. "Ethics, Law Examine Experiments." Part 12,
   "Moral Choices in Contemporary Society." *National Catholic
   Reporter*, April 15, p. 18.
U.S. Congress. Senate. Committee on Labor and Public Welfare.
   Subcommittee on Health. Committee on the Judiciary. Sub-
   committee on Administrative Practice and Procedure. Joint
   Hearing. *Oversight Hearing on Implementation of NIH Guide-
   lines Governing Recombinant DNA Research* (Kennedy Hearings
   1976). Washington: U.S. Government Printing Office.
U.S. Department of Health, Education and Welfare. Public Health
   Service. National Institutes of Health. *Recombinant DNA
   Research, Volume I: Documents Relating to "NIH Guidelines for
   Research Involving Recombinant DNA Molecules" February 1975-
   June 1976* (NIH Documents 1976). Washington: U.S. Government
   Printing Office.

Ethical Theories Underlying the Recombinant DNA Controversy[1]

Mary B. Williams

Freshman Honors Program
University of Delaware
Dover, Delaware

The sources of the disagreements in the recombinant DNA controversy can be separated into three types: scientific, political, and moral. The scientific sources involve disagreements about the amount of damage that would be caused by the production of a particular biohazard, and about the probability that such a production would inadvertently occur; these estimates have been extensively discussed.[2] The political sources involve disagreements

[1] Work on this paper was supported by NSF grant OSS76-16382 from the Ethics and Values in Science and Technology Program.

[2] For example, in the Hearing on Recombinant DNA Experimentation in the City of Cambridge before the Cambridge City Council, July 7, 1976, and June 23, 1976, Ptashne, Singer, King, Wald and others discuss the risk of damage, while Baltimore, Papenheimer, Hyatt, Gilbert, and others discuss the probability of

about the ability of our cultural and political institutions to properly handle the increased power that the research would give us; differing estimates[3] of this ability are a major source of the disagreement in the controversy (particularly in fears about the genetic manipulation of humans), but discussion of this must be deferred to another paper. The moral sources, which are the subject of this paper, involve disagreements about moral theories; I will argue that the opponents of recombinant DNA research are most strongly influenced by a moral theory which focuses on the absolute worth and dignity of each individual human being, while the proponents are most strongly influenced by a theory which focuses on maximizing benefits to humankind.

Before plunging into the discussion of the specific influence of these theories, I must devote a couple of paragraphs to showing how moral values influence conclusions about these scientific matters. First let us consider a rather far-fetched example: Suppose that one of the possible dangers from this research was the creation of a chimera that kills automobiles, and suppose that this car-killing disease would be very infectious and could wipe cars off the face of the earth. The amount of harm caused by this chimera is a function both of the amount of *damage* and of the *value* of the thing damaged. Now, in our culture some people value cars much more than others do, so even if the car-lovers and the car-haters agreed on the probability that it would be produced and on the probability that it would kill all cars, they would disagree on the amount of potential harm that should

benefit. More recently, Cohen (1977) evaluated briefly both the biohazards and the potential benefits, while Roy Curtiss, in a widely circulated letter, examines in detail the probability of producing a pathogenic chimera.

[3]Examples of these different views can be found in the Cambridge hearings cited above.

be put into a cost-benefit analysis to determine if the research
should be continued. Suppose further that there was really no
basis on which to compute these two probabilities, although they
could be roughly estimated in some intuitive way by scientists
who know the basic biology involved; then the scientists would
feel that they could not legitimately give any numbers for the
probabilities. However, if society asked them whether the
research should be continued, the scientists would have to say
whether or not they considered it too dangerous to continue, and
in deciding on the amount of danger they would have to take the
value of cars into consideration. But since the calculations
would all have been made on an intuitive level, neither the public
nor the scientists would know just how the value judgments had
influenced the decision.

Scientists are in this type of situation in the recombinant
DNA controversy; while they have not been able to give simple
factual answers about the probabilities, they have been forced to
give recommendations.[4] There are two lessons to be drawn from
my car-killing example: Firstly, although it would be relatively
easy for scientists to reveal differences in how much they value
cars, it will not be easy for them to reveal differences in how
much they value human lives; in spite of the fact that society
frequently makes decisions (e.g., in deciding how much to spend on
engineering safer highways) on how much a human life is worth, we
look with abhorrence on anyone who openly states that human lives
are not of absolute value. It is hard to expect the scientists
to expose themselves to this abhorrence, and it is possible that,
if they did, society's decision would be based more on an emotional

[4]Note, for example, in the Cambridge hearings, Singer's statement, concerning the probability of the laboratory-produced
organisms escaping into the environment, that the possibility "is
extremely remote. So much so that it is difficult to put any number on it whatsoever" (p. 46).

reaction to this abhorrence than on its real values (as revealed
in its other decisions). The second lesson to be drawn from the
example is that we cannot simply take a vote among scientists
and expect the result to be the same as a vote among the people
would be if they had the same knowledge of the facts as the
scientists had. The proportion of car-lovers and car-haters is
probably not the same among scientists as among the populace, and
it *may* be the case that the values of scientists about the more
important things involved in the recombinant DNA controversy are
not representative of the values of the populace. So, although
it will be hard to get at the scientists' values, it is important
to do so in this controversy.

It is possible (indeed common) for individuals with basically
the same ethical theory to have somewhat different values, but
value disagreements among people with different ethical theories
are likely to be much more serious. So it is important to try to
discover the ethical theories of the protagonists in this controversy. I contend that the proponents of continuing the research
are influenced by a social benefit theory of moral obligation
while the opponents are influenced by an individual worth theory
of moral obligation. A social benefit theory holds that whatever
action maximizes overall benefits is the right action, while an
individual worth theory accepts some moral principles which can
override the result of a cost-benefit analysis. (In the context
of a moral theory 'cost-benefit analysis' refers to an analysis
which balances off benefits against harms.) A quotation from
the philosopher Alan Donagan shows the essence of the difference
between two such theories:

> Utilitarianism conceives our moral obligations to derive
> from the . . . obligation to increase good and diminish
> evil, no matter what must be done to this or that indi-
> vidual. . . . [A Kantian conceives] that moral obliga-
> tions derive from the obligation to respect the

independence and worth of every individual, no matter at
what cost in good forgone and evil accepted. (Donagan,
1974; p. 171)

Using this characterization of utilitarianism, a social benefit
theory, and Kantianism, an individual worth theory, I will show
that these theories impose different moral obligations on their
followers.

Notice that if the 'no matter what' clauses in the descriptions of these principles are left out, the remaining principles
are:

1. We have an obligation to act so as to increase good and
diminish evil.

2. We have an obligation to respect the independence and
worth of every individual.

Most people subscribe to both of these principles, and most of the
time there is no difficulty in doing that because both of these
obligations call for the same action. But sometimes there is a
conflict: one of them tells you that you ought to do action X,
while the other tells you that you ought not to do action X. For
example, suppose you are in an overcrowded lifeboat and it is
clear that all ten persons in the lifeboat will perish unless two
people are thrown overboard. In such a case you must decide
which of these principles is more important, since the utilitarian
principle would obligate you to throw the two overboard, while the
Kantian principle would obligate you not to throw them overboard.

That the utilitarian principle obligates you to throw them
overboard is, I think, clear: the sacrifice of two lives is a
lesser evil than the sacrifice of ten lives. But the obligation
imposed by the Kantian principle is somewhat less clear, and it
may be useful to indicate how the obligation not to throw them

overboard is derived from Kant's own principle, rather than from
the paraphrase given above. Kant's second formulation of the
categorical imperative states: "Act in such a way that you always
treat humanity whether in your own person or in the person of any
other, never simply as a means, but always at the same time as an
end" (Kant, 1785; p. 96). The contemporary Kantian philosopher
Robert Nozick discusses the implication of this formulation for
situations such as the above lifeboat situation:

> Using one of these people for the benefit of others, uses
> him and benefits the others. . . . To use a person in
> this way does not sufficiently respect and take account
> of the fact that he is a separate person, that his is
> the only life he has. *He* does not get some overbalanc-
> ing good from his sacrifice, and no one is entitled to
> force this on him. (Nozick, 1974; p. 33)

Thus the Kantian is obligated not to throw the two overboard.

It is important to recognize that these different answers to
what ought to be done are based neither on differences in amount
of concern for people nor on differences in depth of moral commitment. The utilitarian would regret having to treat the two
unjustly, and the Kantian would regret having to allow all to
perish. The utilitarian would be horrified at the Kantian's
willingness to allow many people to die, while the Kantian would
be horrified at the utilitarian's willingness to commit an unjust
act. But their different actions are based on deep commitments
to different moral theories. Similarly, I contend, the different
answers on recombinant DNA research are based on commitments to
different moral theories. If the proponents of recombinant DNA
research are primarily influenced by a social benefit theory and
the opponents are primarily influenced by an individual worth
theory, then their disagreement can be explained as due to the

fact that the possibility of this research causes a situation in
which these two theories conflict.

I will later sketch the ways in which the conflict arises, but
first let me give another example--one that is much more realistic
and which might help you to see what your own moral theory is.

> During the seige of Leningrad of World War II, several
> thousand people died of starvation. Food was so scarce
> that people were reduced to eating rats, cats, dogs,
> dried glue from furniture joints, and other delicacies
> that might prolong life. All this time, truckloads of
> edible seeds (e.g., wheat) were in storage at the All-
> Union Institute of Plant Industry. (This was the
> Chetverikov collection, at one time the most important
> collection in the world.) These seeds would be of
> major importance in the development of crop plants which
> were needed to prevent future disastrous famines. Accord-
> ing to the best estimates of the scientists, the use of
> these seeds in crop development after the war would pre-
> vent many more deaths than would be saved by using them
> to feed the starving of Leningrad.[5]

Now my question for you is: What do you think the scientists
ought to have done in this situation? Do you think that they
should have saved the people in front of them and hoped, against
all probabilities, that destruction of these seeds would not cause
future disasters? If so, you are primarily influenced by an
individual worth theory.[6] If you think they should have saved the

---

[5]This is a paraphrase of the description given in Harlan,
(1975), p. 621.

[6]One might argue that even Kantians should allow the people
in front of them to starve, since destroying the seed would

seed in order to prevent future famines, then you are primarily
influenced by a social benefit theory. (If you refuse to make a
decision then you should worry about being a moral coward.) The
scientists in fact felt that the seeds were too precious to be
sacrificed even at the cost of human life, and the collection
survived for future use. That this was a decision based on deep
moral commitment rather than callousness is evident from the fact
that the scientists making the decision were themselves in
Leningrad, and starving.

I have used the phrase 'primarily influenced by' because I
think that the positions in this controversy are better described
if the social benefit/individual worth distinction is viewed not
as a dichotomy but rather as defining the two end points of a
continuum. For example, for the Leningrad case suppose that the
expected number of presently starving people who could be saved
were 1,000. Then a person totally committed to an unadulterated
social benefit theory would take the position that if the expected
number of people dying as a result of not saving the seeds were at
least 1,001, then the seeds should be saved. (The expected number
dying would be calculated by taking the estimate of the number
dying and multiplying it by an estimate of the probability that

violate the rights to life of the future famine victims, and the
fact that there would be more future victims would mean that the
violation of their rights would be greater. But Nozick rejects
this argument, asserting (op. cit., p. 32): "Had Kant held this
view, he would have given the second formula of the categorical
imperative as, 'So act as to minimize the use of humanity simply
as a means,' rather than the one he actually used. . . ." So the
Kantian whose principle ranges over both present and future people
has a problem here. If he has only two alternatives (use the
seeds or don't use the seeds) and both involve treating someone as
a mere means, then it would be wrong for him to do either, yet he
must do one or the other.

the estimate is correct; so the possibility that something else
might prevent future famines has already been taken into account.)
A person primarily influenced by a social benefit theory might
hold that the obligation to save them would be overridden only if
the expected number of people saved in the future were at least
1,500, while one primarily influenced by an individual worth
theory might require at least 20,000 future deaths. A person
totally committed to an individual worth theory would require an
infinite number of deaths--that is, he would not allow the obligation to be overridden by *any* number of future deaths.[7]

Now let us return to recombinant DNA. I will take three
issues within this controversy and indicate how the positions
taken on them would be influenced by the acceptance of one of
these ethical theories. The issues are: 1) human genetic engineering; 2) the risk that harmful chimeras will be produced accidentally; and 3) the danger of imposing restrictions on scientific
research.

Both the opponents and the proponents agree that this research
will make available technical knowledge which could be used in
human genetic engineering. The opponents view the possession of
this knowledge as horribly dangerous. (When Ethan Signer was
interviewed on the MacNeil/Lehrer Report, he would speak only
about this danger, though the reporters tried to get him onto
other topics. And the disrupters of the National Academy of
Sciences hearing on recombinant DNA research focused on this
issue, changing "We will not be cloned!" and unfolding a banner

---

[7]Rawls ascribes to Kant the view that, "The value of persons
. . . is beyond all price," (Rawls, 1974; p. 586). And Paton
credits Kant with the "insight" that "the difference between the
good man and the average sensual man is surely that the former
recognizes the infinite and unique value of all moral agents. . ."
(Paton, 1948; p. 174).

with a quotation from Hitler about perfecting the human race.)
The proponents of continuing the research agree that possession
of the knowledge is risky--most of them, I think, believe that
human genetic engineering is undesirable, considering our general
lack of both knowledge and wisdom--but they do not regard this
possibility with the horror that induces the opponents to feel
that it is better to forgo the good which would come from the
research in order to prevent this absolute evil. I contend that
an important source of this horror is the fact that genetic
engineering could be used to violate the autonomy of the individual in a way far more invidious even than psychosurgery: although
psychosurgery, in its worst form, restricts mental freedom, each
use of it is clearly a use of coercion to deprive an individual
of freedom; but genetic engineering, in its worst form, could
make an automaton in human form who had been deprived of autonomy
prior to conception. Even the use of genetic engineering to
remove genetic defects brings objections; Jon Beckwith, who gave
up his research in this field because of his fears of the use to
which it would be put, objects to the underlying assumption that
some humans are defective on individual worth grounds: "The
premise of our interpretation is that a humane society is one in
which the individual is valued by the very fact of being human"
(Beckwith, 1976; p. 55). Thus, since such uses of genetic engineering deny the absolute value of the individual, we should
expect that persons influenced by the individual worth theory will
be much more concerned than others about the risk that the
research will be used in human genetic engineering.

The second issue to be discussed is the risk of harmful
chimera being produced in the research. Notice that in the lifeboat example the Kantian was more concerned with the harm done by
a positive action of his than with the harm done because he took

no action.[8] Because a social benefit theory counts only the consequences (in amount of benefit or harm) of an action, it does not
make such a distinction (although if one consequence of allowing
killing in such cases was an increase in the probability of unjustified killing, that consequence would be counted as part of the
harm resulting from the action); so in the lifeboat situation the
utilitarian simply counted the relative amounts of damage of the
two alternatives, without regard to how the damage was caused.
Similarly a person primarily influenced by a social benefit theory
will judge the advisability of research which may produce dangerous chimera by balancing the possible damage caused by the chimera
against the possible damage caused by the absence of the knowledge
which could be gained by such research. But an individual worth
theory allows the result of a cost-benefit analysis to be overridden by other moral principles; one such principle, familiar
because of its use in the abortion controversy, is that allowing
to die is morally preferable to killing. A generalization of this
principle seems to underlie some of the opposition to recombinant
DNA research--namely, the principle that damage resulting from
natural processes which humans might have prevented is morally
preferable to damage resulting from human action. When someone
using this principle looks at the possible damage caused by
chimeras *which were caused by human action,* he sees the damage as
worse than a similar amount of damage caused by nature. Using

[8]Kantians have not, to my knowledge, argued that the doctrine
that allowing to die is morally preferable to killing is derivable
from the Kantian principle. But their decisions in particular
cases (e.g., that no one ought to be thrown overboard in the lifeboat case, or that the innocent ought not to be executed even to
save other innocent lives) seem to be dependent on an interpretation of the Kantian principle which contains the moral prefera-
bility doctrine.

this principle Chargaff can say: "There is no hurry, there is no hurry at all" (Chargaff, 1976; p. 938). Using the social benefit theory Cohen says:

> . . . humanity continues to be buffeted by ancient and new diseases, and by malnutrition and pollution; recombinant DNA techniques offer a reasonable expectation for a partial solution to some of these problems. Thus, we must ask whether we can afford to allow preoccupation with and conjecture about hazards that are not known to exist, to limit our ability to deal with hazards that do exist. Is there in fact greater risk in proceeding judiciously, or in not proceeding at all? (Cohen, 1977; p. 657)

Clearly, if the proponents of the research based their decision to advocate the research on the result of a cost-benefit analysis, while the decision of the opponents was influenced by this overriding principle, then their decisions would be expected to differ in just the way that they do differ.

The third issue to be discussed is the danger of imposing restrictions on scientific research. Scientists who are particularly worried about this danger contend that any such restriction may put us on the slippery slope leading to more and more restrictions and finally in a virtual end to scientific progress. One can challenge this argument (there is a well known fallacy known as the slippery slope fallacy), but for the purposes of this paper it is more important to discuss a less extreme position which merely contends that the restrictions envisaged would be undesirable because of their own effects on the acquisition of knowledge. Before examining the way in which these two moral theories might influence attitudes about such restrictions, let us consider the difference of opinions about the desirability of imposing restrictions naively as providing a test of moral worth: many of the scientific proponents of few restrictions on scientific research

would personally benefit from fewer restrictions, and some of the
scientific opponents would be personally harmed by greater
restrictions; since personal sacrifice is evidence that one's
motives are good, it is tempting to conclude from these facts
that the opponents must be morally more worthy than the proponents, and that their position must be right; but that conclusion
is valid *only if* the proponents and opponents subscribe to the
same moral principles, so let us now look at how the different
moral theories might influence their position on this issue. A
social benefit theory asserts that everyone is obligated to maximize the total good; there are differences among those holding a
social benefit theory concerning what things (or states of being)
are good, but for many knowledge itself is a good; thus for them
there is, if the gain in knowledge is great enough, a moral obligation to maximize the amount of knowledge even if the actions
necessary to gain that knowledge impose a risk on some human
beings who could not benefit from it. On the other hand, one who
holds an individual worth theory, though he may wish to maximize
knowledge, would be obligated to not impose such a risk. So if I
am right about the moral theories which are influencing the two
sides, the proponents' moral theory would lead them to oppose
restrictions which the opponents' moral theory would lead them to
favor.

Thus it is clear that disagreements among scientists on each
of these three issues involved in the recombinant DNA controversy
could be caused by differences among the scientists in their
moral theories. In my descriptions of consequences of these
theories I have tried to avoid favoritism, but this neutrality
should not be taken to indicate that both theories are right and
the choice between them merely a matter of taste. At least one
of the theories is wrong and the choice between them is a serious
moral choice. However, the purpose of this paper is not to tell
what the proper choice is but rather to point out another level
at which the recombinant DNA discussion should be carried out. If

the disagreement is caused to any significant extent by a disagreement about moral principles, then it cannot be resolved
without explicit discussion of the underlying moral principles.

ACKNOWLEDGMENTS

I would like to thank Dr. Robert Scott for his helpful comments
and criticisms.

REFERENCES

Beckwith, J. 1976. "Social and Political Uses of Genetics in
    the United States: Past and Present." In *Ethical and Scien-
    tific Issues Posed by Human Uses of Molecular Genetics*, 46-56.
    Edited by M. Lappé and R. S. Morrison. *Annals of the New York
    Academy of Sciences, 265.*

Chargaff, E. 1976. Letter. *Science, 192,* 938.

Cohen, S. 1977. "Recombinant DNA: Fact and Fiction." *Science,
    195,* 654-657.

Donagan, A. 1974. "Is There a Credible Form of Utilitarianism?"
    In *Introductory Readings in Ethics,* 165-171. Edited by W. K.
    Frankena and J. T. Granrose. New Jersey: Prentice Hall.

Harlan, J. R. 1975. "Our Vanishing Genetic Resources." *Science,
    188,* 618-621.

Kant, I. 1785. *Groundwork of the Metaphysic of Morals.* Trans-
    lated by H. J. Paton. New York: Harper and Row, 1964.

Nozick, R. 1974. *Anarchy, State, and Utopia.* New York: Basic
    Books.

Paton, H. J. 1948. *The Categorical Imperative.* Chicago: Uni-
    versity of Chicago Press.

Rawls, J. 1974. *The Theory of Justice.* Cambridge: Belknap
    Press, Harvard.

Politics

A Legal Perspective on Recombinant DNA Research

Harold Green

**National Law Center
Georgetown University
Washington, D.C.**

I feel somewhat uncomfortable playing the role of historian, but I do have an historical vignette to relate following up on Dr. Sinsheimer's paper. When President Roosevelt came into office, one of his first actions was to establish a high level National Resources Council to consider how the resources of the United States could be put to work to deal with the depression problems. The Natural Resources Council regarded science and technology as a national resource, and it appointed a high level Subcommittee on Science and Technology--the members of which included some of the most prominent scientists and social scientists in the United States at that time. The report stemming from the work of the Subcommittee, *Technological Trends and National Policy*, is an exceptionally interesting document. The Subcommittee pointed out that we had just completed the first third of the Twentieth Century, a period of unparalleled scientific and technological achievement, and now stood on the threshold of the second third of the Twentieth Century. From that perspective, the Subcommittee undertook to

predict what was likely to happen in science and technology in the
second third of the Twentieth Century. I will not detail for you
everything that the Subcommittee failed to predict in its 1937
report--the jet engine, space exploration, the computer, and atomic
energy. One of the most striking things that they concluded was
that the development of aviation had run its course, and that
future developments were more likely to be in the area of comfort
and safety than in speed.

In the first part of this paper I intend to sketch very briefly the history of events that bring us here today and then offer a
few comments about this history as I see it from my perspective as
a student of the public policy decision-making process. In the
second part, I shall discuss the question of regulating DNA
research. It is my position that the recombinant DNA problem is
not really unique, and that there is firm precedent for the regulation of research that may have significant direct impact on the
public health and safety or on the environment.

HISTORY

The story begins with the 1973 Gordon Conference on Nucleic
Acids where the participants became concerned about the potentialities of recombinant DNA research and addressed an open letter
on the subject to the President of the National Academy of
Sciences and to the National Institute of Medicine. The following
year a committee of the National Academy of Sciences addressed a
letter to *Science* calling for a deferral of certain recombinant
DNA experiments pending an international scientific meeting to
consider what should be done. Significantly, these two events
involved a decision to go public with this problem, to expose the
problem openly.

In February, 1975, the famous Asilomar Conference was held.
There were 134 invited participants including 52 foreign

scientists. Only four of the invited participants were not
scientists, and, strangely, all were lawyers. The Conference
worked out a set of guidelines to match the level of containment
to the level of hazards in various kinds of recombinant experiments. Indeed, the conference reached the conclusion that there
were some experiments that were too hazardous to be performed at
all at that time. Although there were only tentative guidelines,
scientists throughout the world were enjoined to comply with them
until more definitive country-by-country standards were evolved.

The real development of the guidelines was to be by the
National Institutes of Health. At NIH, a Recombinant DNA Advisory
Committee was appointed to work on this problem, and in December,
1975, this Advisory Committee submitted a report to the Director
of NIH. The Director of NIH then decided that there should be a
public meeting of his own Advisory Committee to the Director to
consider the problem. Although the Advisory Committee to the
Director was a rather small group, its membership was increased
for this purpose on an *ad hoc* basis by adding a number of eminent
scientists who were not involved in the recombinant area as well
as a number of non-scientists, including philosophers and lawyers.
Specific invitations to attend, to participate, and to make statements at this public meeting were sent by Dr. Fredrickson to
numerous public interest and environmental groups. Many questions
were raised in this public meeting, a transcript of which was
made, and certain of these questions were referred back to the
Advisory Committee on Recombinant DNA.

Finally, on July 7, 1976, the Director of NIH promulgated the
guidelines that we are discussing today. I should note that there
were a few perturbations in the middle of this process. Senator
Kennedy, who had had a long-standing interest in the biomedical
area, held some hearings on the problem. In the early part of
1976, the principal environmental organizations in the country
suddenly discovered the issue. Some prominent scientists like Dr.
Sinsheimer began expressing serious reservations about recombinant

DNA work. A number of municipalities like Cambridge, Massachusetts
and a number of universities found themselves embroiled in controversies as to whether recombinant DNA molecule work should take
place within their precincts.

While the guidelines were under consideration, some environmental groups who were skilled in working with the National
Environmental Policy Act began telling NIH that it could not
legally issue the guidelines without compliance with NEPA. The
upshot of this was that, with his announcement of issuance of the
guidelines in July, 1976, Dr. Frederickson also announced that NIH
would prepare an environmental impact statement as required by
NEPA. At the same time, Dr. Frederickson noted that there was a
technical deficiency because under NEPA an environmental impact
statement must be published before the federal action is taken,
and in this case was being prepared after the action was taken.
This deficiency was justified on the basis of the desirability of
having the guidelines operative at once instead of deferring them
for another nine to twelve months until the NEPA procedures had
run their course so that such recombinant experiments as were
being performed would be done under appropriate safeguards.

Finally, in March of this year there was a National Academy
of Sciences public forum in Washington that brought together many
of the leading scientists and other figures with respect to this
problem in a wide open atmosphere.

Now I shall offer some comments about this history on the
basis of my own study of decision-making processes, my own personal involvement in this problem (both as one of the four lawyers
who were invited to Asilomar and as a consultant to NIH), and my
own relationships with some of the scientists who were the principal sponsors of the Asilomar conference.

Originally, although I was critical of some parts of this
history from the standpoint of public policy decision-making, I
regarded the events I have described as a paradigm of responsible
public policy decision-making. I still think that may be true,

but I have been beginning to develop significant reservations. I
have no doubt that the sponsors of ths Asilomar conference made a
great effort to expose the entire issue to public scrutiny and to
encourage broad public participation. I believe that they questioned whether recombinant DNA molecule research should proceed,
and that they were prepared to let the political processes work
their will with respect to that issue. They were seriously dis-
appointed when they were not able to engage the active attention
and participation of important segments of the public such as
professional societies of lawyers, economists, philosophers, his-
torians, theologians, and the like. The fact is that the Asilomar
conference was not structured to engage the attention of these
groups. It was essentially a scientific meeting for scientists
with traditional scientific papers, slides, and the like. Frankly,
I do not know why we lawyers were invited to "participate." Personally, although I attended and listened diligently, I did not
have the foggiest idea of what was going on. True, I could sometimes comprehend the general drift of the discussions, and found
some enlightenment listening to lunch, dinner or beer party discussions among the scientists. Curiously, they talked with each
other in plain English in an informal context. Asilomar was
simply not the kind of conference that would produce comprehen-
sible information for the public. Although the conference did
have wide press coverage, with 16 representatives of the media
present with full access to everything, I do not think the scientific tone of the meetings gave the press a real basis for
attracting public interest in the basic ethical and policy issues.
There was, moreover, a serious deficiency in the way the discussion went, a deficiency that probably reflects a fundamental
scientific predisposition. There was virtually no explicit discussion of the potential consequences if the experiments did not
work the way they were expected to. I would have liked to have
heard about the consequences if one of the little organisms
escaped. Are we talking about a little sneezing and minor rashes,

or are we talking about mass cancer? That is the kind of thing
that a layman would have liked to know and could comprehend, but
that was not what the discussion was about. All of the discussion was about how the design of experiments could minimize the
probability that such an event would take place. Once you start
talking about protective mechanisms and probabilities, you are
talking about things that laymen inherently will not understand
and comprehend.

Following the Asilomar conference, as I indicated, the issue
has become very much politicized. The scientists who had spon-
sored the moratorium and Asilomar, and NIH, which favored some
restrictions on research found themselves under severe criticism
from colleagues who regard the restrictions as unnecessary or
excessively severe. Then the same group suddenly found themselves
under attack from the other flank. The environmentalist flank,
aided and abetted by a number of prominent and reputable scientists, have argued that the guidelines are not stringent enough,
and that perhaps recombinant experiments should not be done at
all. As a consequence, NIH and the Asilomar group seems to have,
to use the Lyndon Johnson expression, "hunkered down" in their
bomb shelters and been forced into a defensive posture. Moreover,
as they have become more defensive, they seem to have somewhat
changed their position. The earlier view, to use Watergate termi-
nology, of a "total hangout" to let the public work its will
through the normal political processes, seems to have been replaced
by a "limited hangout." The present tendency seems to be to
encourage vigorous public discussion, but to limit the scope of
the debate. There are assertions that one who is not thoroughly
familiar with all the background is not qualified to speak--or at
least is not qualified to be listened to. There is also an
insistence that the debate proceed entirely on the basis of valid
scientific fact. When a scientist expresses concern about what
may happen under highly pessimistic scenarios, he is challenged
to produce scientific data supporting the pessimistic possibility.

In short, the debate seems to have reverted to the prototypical
case of optimistic resolution of scientific uncertainty. The
fact is, as I understand the situation, that there is a great deal
of uncertainty as to both benefits and risks of recombinant
research. We do not even know with confidence what potential
harmful consequences to look for. Nevertheless, the decision has
been made to proceed with research under strict safeguards that
may or may not be strict enough--no one really knows--and to
start probing those areas of uncertainty in the hope and expectation that everything will work out well.

As I pointed out in my talk at Asilomar, there is a striking
analogy between the recombinant DNA problem and the problem of
nuclear power. In both technologies, you are creating something
that does not exist in nature, something that is potentially
extremely hazardous if it escapes into the environment, and you
are relying entirely upon containment to protect the public health
and safety. In both nuclear power and recombinant DNA technology,
we are going forward on the assumption that human beings are
sufficiently omniscient and infallible to be able to conceive of
everything that can go wrong and to adopt necessary procedures
and safeguards to keep those things from going wrong. And,
finally, there is in both areas a reluctance to discuss the consequences in the event Murphy's Law (if something can go wrong, it
will go wrong) takes over. Nevertheless, to date, the discussion
of the recombinant DNA problem has been many orders of magnitude
more open and candid than the discussion of nuclear power has ever
been. I make this point because of my concern that unless we are
very careful--and by this I mean very open and candid--the entire
DNA issue is likely to become polarized exactly the same way that
the nuclear power issue has been polarized.

REGULATION

Is there anything about recombinant DNA research that should
make it immune from regulation? No, I am happy to say there is
nothing that is unique or novel. There are many areas that we
know have involved restrictions and prohibitions on scientific
research: the entire area of the use of radioactive materials
under the Atomic Energy Act, research with respect to drugs and
chemical food additives, prohibitions and restrictions on human
experimentation, to mention only a few.

Some have suggested that one possibly unusual or unique aspect
about the recombinant DNA molecule problem is that regulation has
been imposed in the absence of any apparent harm. It is true that
our general pattern of regulation of hazardous activities is that
action is taken only when, to use Roy Curtiss' expression, there
is real fear or beyond that a basis for fear. As we well know,
health and safety regulation typically is adopted too late after
many people have been hurt.

There has been a considerable tendency in recent years to
reverse this system. We have had legislation that requires the
pretesting of chemical food additives, new drugs, toxic chemicals,
trace materials, and the like. And, of course, there is one major
example--atomic energy--in which regulation was imposed before
the technology was in existence.

Another question that has been raised is that of the so-called
right to scientific inquiry. A number of scientists have talked
about the right to scientific inquiry as if it were protected by
the Constitution of the United States so that government has no
constitutional power to regulate or prohibit scientific experiments. As a teacher of constitutional law, this strikes me as
great rhetoric in a political argument, but it holds no water as
a matter of constitutional law. First of all, there is nothing in
the Constitution that has ever been explicitly interpreted as
being applicable to any right to scientific inquiry. Because I

believe in scientific and other kinds of freedom, however, I am
prepared to argue that freedom of scientific inquiry is implicit
in the First Amendment guarantee of freedom of speech and press.
I am perfectly willing to assume that it is there, but that does
not get you very far. If one uses the First Amendment as an ana-
logue, First Amendment rights are not absolute. I do not have
the time for a detailed discourse on First Amendment principles,
so I will simply say that First Amendment interpretaions draw a
sharp line of demarcation between pure speech on the one hand and
action on the other hand. Speech is generally constitutionally
protected under the first Amendment, but action is not. I have no
doubt that Congress may not enact a law that interferes with the
right of a scientist to sit in his study or his office and do
calculations, to write things down on paper, to give scientific
papers orally, to publish scientific results, and so on. But when
you get to the point that experiments are performed, you are in
the realm of action and First Amendment protection, if there is
any at all, dwindles in significance. Second, the First Amendment
cases have drawn a distinction between direct and only incidental
restrictions on speech. It may be that the City of Cambridge,
for example, might adopt health, safety, or environmental regulations to restrict the conduct of recombinant DNA experiments within the city limits. Although such restrictions might have some
effect on the freedom of scientific inquiry, they are really
adopted to protect the environment and are only incidentally a
regulation of speech. Some of my scientist friends would not dispute my analysis but would go on to argue that Cambridge has the
burden of proof to show that unregulated experiments will in fact
jeopardize health, safety, or the environment. That does not cut
any constitutional ice either. People who make that kind of
argument are harkening back to the bad old days which many of you
will remember--the early years of the New Deal and before when
the Supreme Court held unconstitutional all kinds of forward-
looking regulatory legislation because it thought the legislation

was unnecessary or undesirable. Ever since the late 1930s, however, the Supreme Court has consistently followed the policy of
not interfering with the judgments of state legislatures and the
Congress as to whether a particular regulation is necessary,
desirable, or wise. There will be judicial interference with the
will of the legislature only under two circumstances: where there
is no rational basis for the regulation, or where the regulation
violates some explicit constitutional guarantee. Neither of these
circumstances is present in connection with the recombinant DNA
molecule problem. Therefore, my bottom-line judgment is that it
is constitutional for Cambridge, for Georgia, and for the United
States Government to regulate, and indeed to prohibit, recombinant
DNA molecule research if it sees fit.

The next question is who should do the regulation? Mr. Clem
makes a dramatic appeal for separating regulation from promotion.
I might say that from the very outset NIH has said that it should
not be the regulatory agency, and under the proposed legislation
it will not be. I question, however, whether we advance the cause
of regulation very far by making some other arm of HEW the regulatory agency. If we are serious about separating regulation from
promotion, the regulation ought to be the responsibility of some
outside agency, one that is completely outside the ambit of HEW.
However, as a long-time watcher of the atomic energy program, I
am somewhat skeptical about Mr. Clem's point. It is clear to me
that the problem with the Atomic Energy Commission was not the
combined functions of regulation and promotion in the same agency.
The problems were much more deep-seated than that. The Energy
Reorganization Act of 1974 abolished the Atomic Energy Commission
with AEC's promotional functions being assigned to ERDA and its
licensing and regulatory functions to the Nuclear Regulatory Commission. This was a cruel hoax, because the fact of the matter is
the Nuclear Regulatory Commission today is far more promotional in
its posture than the AEC ever was. Accordingly, it follows that I

do not think the locus of regulatory authority is as important as the spirit in which the authority is exercised.

A second question relating to who does the regulating is whether the area should be preempted by the Federal Government or whether there should be latitude for state regulation. If the federal regulatory agency so conducts itself as to enjoy the full confidence of the public, I would prefer federal preemption. On the other hand, if the regulatory agency lacks such confidence, then there should be latitude for local governments to regulate because the public would probably have greater confidence in local regulation. Another reason for federal rather than state or local regulation is the simple fact that we have only one world and hazardous experiments in Alabama or Oregon could endanger the health and the environment of Georgia to the same extent as recombinant experiments performed right here in Athens. I doubt that anyone could feel much more secure if the hazards are real in having stringent local regulation where regulation in neighboring jurisdictions is much more relaxed. Indeed, this last point also points up the necessity for dealing with the problem of regulation on an international level, because experiments that are performed in the Soviet Union or China could adversely affect the health and environment of Georgia and South Carolina.

I would hope that we would try to avoid being in the box of doing wrong things just because the Russians, or the French, or the Swiss are doing them. One of the things that was uppermost in the mind of NIH when it issued the Guidelines without waiting for compliance with NEPA was the importance that the United States take world leadership in a moral sense in establishing guidelines as quickly as possible in the hope that the rest of the world would follow suit. It was also reasoned that the NIH Guidelines would have greater influence than if NIH policy prohibited recombinant research altogether because there are some countries on this planet that simply will not prohibit this kind of research under any circumstances.

Finally, a recurrent theme at Asilomar was the almost paranoid concern about regulation. Scientists got up and said, "If you publicize this matter too much you are going to get regulation"; and others stood up and said, "If you don't publicize it enough you are going to get regulation." Regulation was the big bad wolf, and the worst of the big bad wolves were those pieces of regulation that, as many of the scientists said, were in the form of a detailed code. The prevailing view was that general regulation was less obnoxious to scientists than specific regulation. It is interesting to note in this connection that the biomedical community is much less knowledgeable about regulation than are other parts of the scientific community. Unlike the harder sciences--like physics and chemistry--there has been relatively little regulation impacting on biomedical scientists. The biomedical sciences have largely been left free to regulate themselves.

The point I would like to leave you with is simply that the recombinant DNA molecule problem is in no sense unique. There is no reason why it should not be regulated to the same extent as other activities that raise a threat of injury to the health and safety of the public or to the environment. And, most important, despite the obviously good intentions and obviously good works done by the recombinant DNA researchers, there is no justification for treating their arena as a sacred preserve immune to regulation by government.

In closing, I would like to say that it would be a serious mistake for the scientific community to attempt to protect the public from itself, to try to protect the public against emotional or irrational reactions to information that might thwart the development of a beneficial technology. After all, the real issue is what public policy should be. What benefits do the public want and what costs are the public willing to pay in order to have those benefits? In a democracy, the public is entitled to choose for itself, rationally or irrationally, emotionally or calmly,

what benefits it wants from its government and what price is it
prepared to pay. Our challenge is really to perfect the democratic
political process to facilitate a proper and constructive role for
the public. Because I have faith in the democratic system, I
firmly believe that if we give all the facts to the public, a
total hangout scenarios and all, the public is sufficiently wise
and mature to make a sound decision.

REFERENCES

*Technological Trends and National Policy: Including the Social
    Implications of New Inventions.* Report of the Subcommittee on
    Technology to the National Resources Committee. Washington,
    D.C.: U.S. Government Printing Office, June, 1937.

Regulation of Recombinant DNA Research

Susan G. Hadden

Southern Center for Studies in Public Policy
Clark College
Atlanta, Georgia

The regulation of research is the central issue of the whole debate on recombinant DNA. Whether to regulate, how much to regulate, and the form of regulatory institutions are questions whose answers will finally determine the course of recombinant DNA research in this country. Scientists are especially worried that research may be over-regulated: in Roy Curtiss' words, "regulated out of blind fear." Their reaction has been to try to limit regulation to the minimum, calling for loose and flexible guidelines that will be responsive to the changes in the state of knowledge.

I will argue here that scientists' interests are not protected by loose regulation, but rather that both science and society are best served by establishing careful guidelines on a variety of issues beyond physical and biological containment in any legislation that is drawn up. The previous history of regulatory agencies given a loose mandate only to protect the public

207

interest shows that these agencies are often "captured" by the
very groups they were expected to regulate, which may in turn
cause harsh public outcries against both agency and clientele and
result in over-restrictive legislation. Such loose regulation as
many scientists seek may well give them strong control now but
extra restrictions later. Although the NIH Guidelines, which many
scientists say they are willing to accept, appear to be specific,
in fact they give considerable scope for judgments by experts;
for example, in assigning particular experiments to containment
categories. This is precisely the kind of situation that tends to
foster clientele capture.

Furthermore, an examination of the literature on the use of
expertise in policy-making reveals an anomalous situation: while
there is a strong tendency to delegate too much power to experts,
at the same time there is an inability to make use of expert
advice except in time of crisis. I will argue here that the
regulation of recombinant DNA research offers a unique opportunity
to evolve a model for ensuring that society's interests are not
compromised by advances in research at the same time that scientists are given a stronger role in the determination of public
policy. It is clear that a great number of technical advances in
the near future will pose problems similar to those posed by
recombinant DNA research; therefore it is especially important to
evolve a model for regulation and oversight that will obviate the
necessity for repeatedly debating the same issues.

REGULATORY AGENCIES AND THE PUBLIC INTEREST

Most federal regulatory agencies are an outgrowth of the Progressive and reform eras of the early twentieth century when
public concern over the special costs of economic activity manifested itself. Regulatory agencies were embodiments of a thesis
that apolitical experts could determine the public interest and

formulate rules that would advance it. Thus most regulatory
agencies received a delegation of power from Congress that permitted them not only to implement the intent of the legislature
but in fact to determine that intent.[1]

From this vague and broad mandate, it is possible to infer
why regulatory agencies gained a reputation for being "captured"
by their clientele, that is, becoming over time more responsive
to those they were supposed to be regulating than to the general
public. First, only the regulated industries had expertise to
match that of the commissioners of the regulatory bodies; apolitical experts, like others, find it easier to communicate among
themselves. Similarly, commissioners know well that as their
terms expire the only jobs that require their painfully-acquired
expertise are in the corresponding industry. Second, the regulated industries have a large stake in the agencies' decisions,
standing to spend, for example, several millions of dollars on
pollution control devices, while beneficiaries of strict regulation gain or lose but little *as individuals* and are, moreover,
most often poorly organized.[2] Thus industries have found it
worthwhile to organize and either to contest restrictive actions
of regulatory agencies or, more often, to seek compromises with
them before decisions are publicized. One cannot be surprised
that an agency with a mandate only to further the public interest
seeks to further the combined interests of itself and of its most
vocal publics.

This pattern of clientele capture applies more accurately to
policies of the pure regulatory rather than of the distributive

[1]Marver Bernstein, 1955; 101, 114. For a more current bibliography on regulation and a useful analysis in itself, see Roger G.
Noll (1971). Also see Mark F. Massell (1961).

[2]This point has often been made. See especially Mancur Olson
(1965), (1975).

types. These are two of four categories suggested by Lowi in
elaborating his innovative idea that the pattern of politics is
determined in large part by the type of policy.[3] Distributive
policies allocate specific goods to groups, and the receipt of the
good by one beneficiary does not affect its receipt by any others.
Rivers and harbors bills, procurement of defense materials, and
research contracts are common examples of distributive policies.
Regulatory policies, on the other hand, directly pit some groups
(the regulated) against others (often "the public") and involve
the making of rules to govern the behavior of the affected groups.
Regulatory policies usually involve the creation of regulatory
agencies which in turn devise highly technical specifications or
rules that are designed to ensure public safety or efficient
economic practices. Typical regulatory agencies include the
Interstate Commerce Commission (ICC), the Food and Drug Administration (FDA), and the Environmental Protection Agency (EPA).

I believe that the control of recombinant DNA will fall into
the area of regulatory policy. All of the rules proposed so far--
the NIH Guidelines, at least two bills proposed in Congress, the
University of Michigan regulations, the Cambridge regulations, and
the laws proposed for California and New York, among others[4]--are
all designed to control actions of a well-defined group by licensing and oversight. Almost all these bodies specify a governmental
(or institutional) obligation to protect the welfare of the public

[3]Theodore Lowi (1964). This typology is further elaborated
in several of Lowi's later writings and in Robert Salisbury (1968).
Also see Sabatier (1975).

[4]In order, Federal Register 41(176)(9 Sept. 1976), pp. 38426-
38483; 95th Congress 1st Session H.R. 4749 and H.R. 3592; Report
Genetics and Oncology Program (Committee B), University of Michigan, March 1976; *Science*, 195 (21 January 1977), pp. 268-269;
California Bill #757; an act to amend the public health law (1977).

as justification for their rules, and imply, at least, a continuum
of tradeoffs between new knowledge or its fruits and public
safety. These are clearly characteristics of a regulatory policy.

What then can we expect to happen? If the granting of licenses to use airwaves or the setting of train fees were considered
subjects so technical that they required experts for policymakers,
then recombinant DNA must require expertise increased exponen-
tially. But all our past experience has shown that experts in
regulatory agencies are especially subject to clientele capture,
and to a tendency that grows over time to equate the public
interest with that of the constituencies they are most like and
most sympathetic with. The risk of equating minimal regulation
with the public interest is maximized in the case of recombinant
DNA, because the stakes will include not only practical benefits
but also the protection of a fundamental principle--freedom to pursue knowledge.

Experience thus suggests that regulatory agencies established
with a broad mandate to further the public interest are more
likely to be captured than those established by legislation that
is specific about goals and purposes and provides policy guidelines within which the regulatory body must work. Most of the
legislation proposed to regulate recombinant DNA research is of
the broad mandate type, with the exception that provisions are
made for physical and biological containment similar to the NIH
Guidelines. Even these are vague in some ways, as I shall point
out below. Much legislation relies upon a central commission to
draw up all rules that it deems necessary beyond those of the NIH
and to implement them with the aid of decentralized local "bioethics committees." If we are to avoid delegating more policy-
making power than we would prefer, however, legislation should be
as specific as possible on all topics that affect the direction of
public policy towards research. I will elaborate on these topics
below.

AREAS OF REGULATION

Up to this point I have stressed those experiences with experts in policy-making that have helped to undermine confidence in their disinterestedness and concern for "the public interest." In discussing the areas for which specific guidelines need to be included in any legislation, I will point out that additional expert advice is required in order to draft a law that is enforceable, efficient, and responsive to the existing state of knowledge.

I have commented on the tendency of experts to talk to each other. This holds true as well in the case of regulating recombinant DNA research. The Asilomar conference was an attempt at self-regulation, which is admirable but does little to alleviate public suspicions. Even now, it is difficult to obtain any literature detailing the evidence that supports the growing conviction of most researchers in the area that recombinant DNA research is generally safe.[5] It is essential that scientists increase their efforts to communicate with the public, both to increase the probability that informed and sensible public policy will be formulated and to prevent legislation out of "blind fear." The public has a right at least to try to understand even the most technical bases for the recommendations of experts.

Above I stated that regulatory legislation should be as specific as possible. There are several areas which are not addressed by most of the legislation proposed to date, especially state legislation. (The bills of Congressmen Metzenbaum and Rogers as well as the "administration bill" propose studies of these areas to be embodied in future legislation.)[6] These areas are:

[5] Curtiss' letter to Frederickson, April 12, 1977, is a notable exception.

[6] S.B. 945, H.R. 4759 and S.B. 1217, respectively (1977).

1. what types of research will be permitted or encouraged;

2. shall licenses be issued for persons or for institutions, or both;

3. what sorts of enforcement procedures will be employed.

In addition, the NIH Guidelines, which give the appearance of specificity, leave some matters unsettled. They will be treated as a fourth category for comment.

Each of these questions entails a number of issues. I will only be able to remark on a very few of them, but I hope that they will stimulate further discussion and, finally, more specific legislation. I will treat each question in turn.

1. What Types of Research Will Be Permitted or Encouraged?

There appears to be general agreement about at least temporary restrictions on research involving pathogenic organisms, the creation of toxins, the transfer of drug resistance, or the spreading of new organisms. Beyond this, however, the subject remains open.

Many analysts have answered the question by recourse to "benefit-cost" analysis. This criterion compares, as its name suggests, gains with costs or risks and accepts a course of action if the former outweighs the latter. Originally devised as a technique to enable the Army Corps of Engineers to evaluate federal expenditures on navigation, cost-benefit analysis has been extended to cover virtually every government and corporate decision made.[7]

---

[7] See almost any public finance text; for example, Musgrave and Musgrave, 1973; 155 ff. Also see F. J. Mishan (1973).

Risk-benefit analysis, a variation, is used in the bio-medical field. It began as a means of calculating the safety of drugs for humans, in which advantages from treatment could be compared with unpleasant side effects and the disadvantages of not using the drug. Such calculations are difficult, but have the advantage that the factors being measured are in comparable units. Benefit-risk calculations that compare health with monetary costs are more difficult, although they have been attempted; such calculations are implicit in decisions about kidney machines, for example. However, benefit-risk calculations that compare producer benefits (in dollars or knowledge) against consumer risks (health) would seem to have reached a point at which too many incommensurables are being balanced. This is especially true when the risks are assumed unknowingly and without choice. This latter type of calculation is the one implicit in the analysis of recombinant DNA research.

To take this point, which I believe to be a most important one, a bit further, several commentators have minimized the importance of considering the risks of recombinant DNA research, which are probably small and clearly unknown. Nothing ventured, nothing gained, they say. On the other hand, these commentators hold out visions of cheaply produced hormones, and solutions to the food problem. Here are measurable benefits.[8] Clearly, these writers are saying, benefits outweigh risks.

In a slightly different way from that prevailing in regulatory agencies, these experts are asking to be allowed to make policy by default. They are saying in effect that laymen cannot in fact judge the risks but that experts should be allowed to do so. There is no question that expertise does entail some sort of

[8]For example, see Stanley N. Cohen (1977). Cohen notes that the unprecedented attempt by scientists to take precautions against unknown hazards has created the (incorrect) impression that such hazards are large.

intuitive understanding that cannot be attributed to the most
educated lay person.[9]   But since risk-benefit analysis of recombinant DNA research involves the balancing of unknown costs with
immeasurable benefits, the skills of the expert are not entirely
sufficient. It is more appropriate that risks, however, small or
remote-seeming, be enumerated and explained so that laymen will
have a chance to balance them with benefits that are, after all,
only projected. It is especially appropriate when the intuitions
of experts lead them to opposite conclusions, as in the case of
recombinant DNA.

Another method that has been frequently mentioned at this
conference is "case-by-case" analysis (most often of the risks and
benefits) which, it is argued, will provide maximum flexibility
and minimum unnecessary control. Again, I must argue against a
course that seems to have wide appeal in the research community.
Case-by-case analysis is extremely wasteful of resources, causing
the same questions to be debated over and over again. More
important, because of the reduced social stakes in any one
decision, there is a tendency to allow single case debates to be
dominated by those with a more explicit interest, *i.e.*, a tendency
implicitly to delegate power to experts, in this case, DNA
researchers. Still another danger is that incremental change
usually appears harmless and so is permitted. This often results
in the familiar situation in which it is argued "we've already
spent millions of dollars on the research so you may as well let
us go ahead and develop the thing rather than wasting all that

[9]For similar sentiments on a related issue, see Hubert Block:
"If you believe, as I do, that it is possible to get a pretty good
feeling for a drug relatively early in the game, it seems wasteful
to spend years getting more data just so that people can have a
spurious sense of confidence in what they know and do not know
about a drug" (Block, 1973; p. 263).

money." Novick's reference to the SST is a case in point; at
each step this argument was used to gain Congressional funding for
further work.

Some participants at this conference have also suggested that
the distinction between basic and applied research is relevant to
the debate about regulation. They believe that basic research,
which is small in scale and is probably conducted by trained
laboratory workers, poses very little threat to public health or
the environment, while applications, which are of necessity larger
scale and may be conducted in the conditions of the assembly line,
may well pose such threats. Without exploring the validity of
this claim, its is surely worthwhile to note that the distinction
between the research and its applications is very difficult to
describe or to enforce; if past trends are any indication, the
size and content of basic experiments will gradually increase so
as to be indistinguishable from experiments leading to the application of the technology.

What sort of criterion, then would be appropriate for select-
ing types of research to license? Implicit in the present and
proposed guidelines is that safety is of overriding importance.
Within this constraint, presumably research that is most likely to
answer important questions will be chosen, but as of now this
aspect will *not* be the subject of regulations. Until some further
experience is gained, the emphasis seems appropriate. Unfortunately, it will serve especially to discourage work by industry,
where dollar benefits must eventually be obtained. However, the
costs of the delay may be more than balanced by the sense of
public confidence in results obtained by research that is truly
publicly sanctioned.

Another criterion which we often disparage but is of considerable importance is the political goal. It is easy to forget that
only by fulfilling the desires of the people can politicians
enable our political system to continue; and it is a political
system that has bestowed more freedom and riches on science than

any other.   Thus popular fear of research will indeed result in
regulation; the scientist's duty to himself and to society is
again to publicize what is known so that politically acceptable
decisions will be ones more consonant with minimum direct control.

2. *Shall People Or Institutions Be Licensed?*

   This area is addressed by much of the proposed legislation,
but again without specifically making policy choices on some
questions. If people are licensed, on what basis? Can they take
their licenses with them? Is an institution licensed because of
the presence of licensed researchers (as well as of appropriate
containment facilities)?

   The personal licensing question was raised emphatically in the
New York State study (see New York Report). There it was suggested
that adherence to strict containment practices was only as good as
the practitioners; only a small proportion of researchers, to say
nothing of technicians and janitors, are properly schooled in the
laboratory techniques of medical microbiology and in the disposal
of wastes. A proposal has been offered that would require each
licensed recombinant DNA worker to have taken and passed an
appropriate short course in lab practice. While this might pose
enforcement problems, experience with Red Cross swimming certification and certification of different levels of ambulance technicians suggests that a decentralized system of training courses
and licensing could be implemented successfully at low cost.

   Institutional licensing at present is limited to examination
of the physical containment criteria. Some commentators have suggested restricting the number of licenses or limiting them to
facilities in unpopulated areas. Others suggest continued inspection for the presence of insects, waste disposal facilities, etc.
(see New York Report). The question of numbers of institutions

licensed must be addressed explicitly, rather than resting upon
implicit assumptions of the NIH Guidelines. Although a limited
number of centralized research sites would facilitate enforcement,
the intellectual checks and balances of the present decentralized
system would be sacrificed. The costs of terminating research
in progress must also be considered.

3. *What Sorts Of Enforcement Procedures*
    Should Be Used?

    Present regulations and proposed regulations are almost unani-
mous in the procedure chosen. They rely upon some central regulatory policy-making committee supplemented with decentralized
local bioethics committees. While there is a clear effort to
include "ordinary citizens" and disinterested parties in the
centralized bodies, there is little attempt to expand the local
bodies in a similar fashion. Decentralized bodies are often
especially liable to cooptation (as in the example of agricultural
extension agents and their general responsiveness to local influ-
entials), and need to have citizen members as well. The most
critical aspect of enforcement will be ensuring that individual
laboratories are practicing proper procedures.
    The decentralized nature of present-day research facilities
renders this problem extremely difficult, and suggests the special
importance of the local bioethics committees. For example, requiring frequent testing of cultures to ensure that they have not
changed character--weakened strains reverting to type, for
example--would seem to be a good rule but it would be virtually
impossible to enforce if researchers chose to ignore it. We know
that adherence to regulations will be strongest where incentives--
carrots rather than sticks--are greatest. If researchers are not
convinced that any risk exists, enforcement will be extremely
difficult, and enforcement bureaucracies will proliferate faster

than the rules they are enforcing. Formal centralized policing
of research facilities which is very expensive must thus be concentrated upon those areas where outside persons may unknowingly be
exposed to risk; for example, through waste disposal. These kinds
of enforcement rules have the advantage of being more easily
measured than those concerning laboratory procedures.

4. Are the NIH Guidelines Of An Acceptable Degree Of Specificity?

The NIH guidelines, which are incorporated in much of the proposed legislation and which are widely accepted in the research
community, define levels of physical and biological containment,
which are quite specific. They would appear to be an exception to
my criticism that too much of the legislation is vague or broad.
However, specific rules of the NIH Guidelines only become effective after a series of determinations is made about the character
of a particular experiment--and those determinations are to be
made by scientists involved in the research. Thus, the NIH Guidelines create a situation similar to that which has led to clientele
capture in other agencies, i.e., the setting of criteria for certain types of control, with the classifications of cases left to
those most knowledgeable and interested. It is clear that scientists will tend to classify their experiments into the category
requiring the smallest amount of containment, and that the regulatory body will accept these classifications because they will
require less enforcement and will cost the scientific researchers
and their patron, the federal government, less money.

This problem is a difficult one to circumvent. The solution
does not lie in the promulgation of thousands of minutely detailed
regulations. As Novick has pointed out elsewhere, the NIH Guidelines on physical containment fill a 93-page booklet which no one

can remember or fulfill.[10] Yet even that booklet missed some
critical points, such as the treatment of aerosol sprays also
mentioned by Novick. John Richards' article in this volume
addresses other such lacks.

The solution to the problem does lie in stating clearly and
carefully what policy aims are and how priorities are ranked so
that the guidelines that are established actually *guide*. For
example, a policy statement that guides choice might be "in all
instances in which there are differences of opinion about the
hazardousness of an experiment, the highest relevant degree of
containment shall be chosen." Another might be "All researchers
must be certified as being trained in microbiological techniques."
The establishment of such policies requires that society conscious-
ly choose and order its goals, which cannot be done if scientists
are the only ones formulating policy or implementing broad guidelines. Opening the debate to all is politically a difficult
course, but the alternative is loss of public control in yet
another "regulated" area.

Two further enforcement issues I would just like to mention
are whether local bodies will be allowed to enforce stricter
regulations than those that apply at the more centralized level,
and what sorts of controls can be enforced on industry and
especially on problems associated with patents.

A THREE-TIER REGULATORY APPROACH

Some of these problems would be addressed by a three-tier
regulatory approach. A central policy-making body at the state,
or preferably, national level, and the institutional bioethics

[10]The testimony given in U.S. House of Representatives (Hearings), p. 148.

committee envisioned by most of the present legislation would be
supplemented by a small state or multi-county body charged with
formal enforcement and some policy-making. While most policy
questions are national, more localized bodies may wish to ensure
that distribution of facilities, or encourage research with a
particular practical focus. Enforcement of waste-water standards
and building codes fits appropriately with services already provided at the local level. The professionally-oriented but diverse
bioethics committee would remain the primary means by which
laboratory practices are controlled. This system would still provide considerable flexibility, allowing for local variations in
needs and goals. In order to ensure further that regulations are
relaxed or strengthened as experience dictates, it is important
that the system not be over-institutionalized. For this reason,
persons at all levels of the regulatory structure should be given
very limited terms of office.

A final comment is necessary here. We have available one
further regulatory mechanism which is in some ways the most powerful of all--money. Although research will continue under any circumstances, withdrawal of funds will surely slow progress. It
would be possible to decide not to appropriate any funds for
recombinant DNA research (or research on any other topic of concern). Another area for advice from experts but not control by
them is the extent of funding that would be appropriate as well
as the particular topics that most need exploring. It is most
important that debates over the extent of funding not be secret,
again so that the public is not presented with a *fait accompli* as
in the case of the SST.

CONCLUSION

In sum, a review of the use of expert advice in policy-making
suggests an anomalous situation in which expert advice is

underutilized at the same time that too much power is delegated to
experts to regulate their own affairs. The etiology of other
regulatory agencies gives us cause to believe that the kind of
broad-scale, vague mandate to protect the public interest which is
proposed by several laws to be entrusted to a regulatory agency for
recombinant DNA research will be exploited to the benefit of
researchers. Therefore, I have argued that any legislation must
include rather specific guidelines addressing questions such as
the nature of research to be licensed, the licensing of institutions and of individuals, and the type of enforcement that can
reasonably be expected to be fruitful.

These are issues which will require the careful use of expert
advice, however, so that proposed regulations are minimally
restrictive. While some restriction is needed in order to ensure
public safety, *minimal* restrictions are desirable to keep enforcement down, to ensure continuing freedom of research, and to make
available such benefits as can be gained. Scientists have an
obligation to publicize their knowledge so that society is not
forced to delegate decisions about research to experts whose
assessments of risks and benefits may not correspond to that of
the public at large.

Debate on the regulation of recombinant DNA research offers us
an opportunity to create a model for future relationships between
science, government, and the public. Governments have tended to
make most use of scientists' advice at times of crisis. The
present case is nearly unique in that discussion has been generated
(at the insistance of scientists themselves) before any crisis
exists. If we can institutionalize a means of regulating research
that will not only ensure the safety of non-researchers but also
will serve to alert them to future developments and their possible
consequences, then we will have served science as well as the
public. Government will have a way of using science advice early
in the policy process when that advice has the greatest effect,
rather than waiting until a crisis exists; science will thus have

a more influential role in policy-making, precisely because it
will be well-defined and regulated. The three-tier approach I
have outlined is suited to this broader task, since regulation
itself is carried out by lower-level bodies while discussion of
policy issues--including new developments in technical capa-
bility--is carried out by the central body composed of a wide
variety of scientists and citizens.

Don Price once noted that the scientists' freedom from politics
depends on the assumption that science cannot authoritatively
define social ends.[11] Let us not be seduced by the difficulty of
the subject into once again offering experts a mandate to define
social ends, especially in an area that may have very broad social
and political implications. The result of doing this before has
been to discredit the scientist and, as Price predicts, to saddle
him with often unnecessary restrictions. Let us rather devise
legislation now that utilizes to the fullest that which scientists
do know to construct rules that are based on understanding and
not fear and which will therefore leave the scientist free to work
within more limited and well-defined constraints. Both the
scientist and the public will benefit.

REFERENCES

Bernstein, Marver. 1955. *Regulating Business by Independent
     Commission.* Princeton: Princeton University Press.
Block, Hubert. 1973. "Towards Better Systems of Drug Regula-
     tion." *Regulating New Drugs.* Edited by Richard L. Landau.
     Chicago: University of Chicago, 243-265.
Cohen, Stanley N. 1977. "Recombinant DNA: Fact and Fiction."
     Science, 195, (February), 654-657.

[11] Don Price (1965).

Lowi, Theodore. 1964   "American Business, Public Policy, Case-
    Studies, and Political Theory." *World Politics, 16* (July),
    677-715.

Massell, Mark F. 1961.   "The Regulatory Process." *Law and
    Contemporary Problems, 26,* 191.

Mishan, E. J.   1973. *Economics for Social Decisions.* New York:
    Praeger.

Musgrace, Richard A., and Musgrave, Peggy G. 1973.   *Public
    Finance in Theory and Practice.* New York: McGraw Hill, 1976.

Noll, Roger G.   1971. *Reforming Regulation.* Washington: Brook-
    ings Institution.

Olson, Mancur. 1965. *The Logic of Collective Action.* New York:
    Schocken Books, 1968.

Price, Don. 1965. *The Scientific Estate.* Cambridge: Harvard
    University Press.

Recombinant DNA Research Act of 1977   (Hearings). Hearings before
    the Subcommittee on Health and the Environment of the Committee
    on Interstate and Foreign Commerce, House of Representatives,
    March 15-17, 1977. Washington: U.S. Government Printing
    Office.

*Report and Recommendations of the New York State Attorney General
    on Recombinant DNA Research*   (New York Report). February
    8, 1977.

Sabatier, Paul. 1975.   "Social Movements and Regulatory Agencies:
    Toward a More Adequate--and Less Pessimistic Theory of
    'Clientele Capture'." *Policy Science, 6,* 306-310.

Salisbury, Robert. 1968.   "The Analysis of Public Policy: A
    Search for Theories and Roles." *Political Science and Public
    Policy.* Edited by Austin Ranney. Chicago: Marcham Publishing,
    151-178.

Should Recombinant DNA Research be Regulated?

Tom L. Beauchamp

Department of Philosophy and
The Kennedy Institute
Center for Bioethics
Georgetown University
Washington, D.C.

In this paper I discuss whether it would be justified to regulate recombinant DNA research. This is a question about regulation, not prohibition, and I do not see any a priori legal or moral barrier to such regulation. Scientists are accountable for public harms as much as anyone else, and their research carries no unique constitutional protection. I can even envision several reasons why legislation might appear favorable from the perspective of the scientific community itself. The regulation could take the form of coercive state regulation or could be a matter of self-regulation by scientist's doing the research. I shall be discussing which alternative is preferable.

The deeper philosophical problem from which this particular controversy about regulation stems has to do with the justification of limiting liberty. Regulatory laws have two sides. By ensuring rights, liberty or protection from risk of harm to one set of

225

persons, a law may restrict the liberty of others. Here emerges
the struggle over regulating scientific research: the freedom of
scientists might be seriously curtailed in order that nonscientists
be protected against risk. It is often said that we willingly
trade some liberties either for the insurance of other liberties
or for some form of protection by the state. But in the present
case it is the willingness of nonscientists to legislate protec-
tions which must be contrasted to the unwillingness of scientists
who do the research. The acceptability of any such liberty-
limiting law will ultimately depend upon the adequacy of the justification offered for it; and when an adequate justification is not
forthcoming, the law may become an instrument of oppression. Laws
governing DNA research clearly could be oppressive for some. The
question is whether they are justified.

I shall first sketch those schemes of legislation presently
contemplated at the federal level (Sec. I). The most plausible
justifications for such legislation will then be discussed (Sec.
II). Finally, I shall inquire as to the adequacy of these justifications, including a treatment of the question whether self-
regulation is better than federal regulation (Sec. III).

Senator Edward Kennedy once stated his conviction that there
should be federal regulation of DNA research as follows:

To my mind this unprecedented dialogue between the scientists
and citizens must be built into the regulatory process.
Because the risks, if there are any, will involve scientists
and lay persons alike, decisions as to how to contain them
must be made cooperatively. I believe a national commission,

with a majority of nonscientists, would be the best vehicle to regulate this research.[1]

Senator Kennedy, who in later months drew back from active support of Senate Legislation,[2] is here voicing the widely held view that regulation at the federal level through statutory law is the appropriate legislative response to the DNA problem. Many senators and the current administration in the White House have expressed similar convictions. Either separately or in combination they have proposed extensive and detailed legislation, some examples of whose main features may be summarized as follows:[3]

1. There should be national standards, with an option for stricter local standards (upon approval by the Secretary of DHEW).

2. A National Commission should be created to study the problem. It should contain both scientific and nonscientific specialties, should study the role of risk/benefit criteria, and should propose mechanisms for evaluating and monitoring the research.

3. The regulation should govern all recombinant DNA research, whether it be done by industry, universities, government or any other person or institution.

4. The number of licensed facilities should be limited, though the numbers should be more than one. Licensing should be rigorous, and projects undertaken should all be registered.

[1] Opening Statement of Senator Edward M. Kennedy, Chairman of the Senate Subcommittee on Health and Scientific Research, at a Legislative Hearing on Regulation of Recombinant DNA research. Dirksen Office Building (April 6, 1977).

[2] Wade (1976) and SGR (1977).

[3] Cf. 95th Congress, 1st Session: S. 621, S. 945, S. 1217 and similar legislation in the House of Representatives.

5. An international conference on legislative problems
should be funded.

6. New laboratory controls beyond the present NIH Guidelines
are probably not needed and could well be too confusing and
oppressive. (Federal regulations would in effect convert the NIH
Guidelines to standards for regulatory purposes.)

7. Mechanisms for inspections and the monitoring of the
research are needed--perhaps through the offices of the Center for
Disease Control.

There are two main alternatives to such federal regulation.
The first alternative is simply to have no regulative devices
whatsoever, and the second is to have self-regulative standards
promulgated by the scientific community, probably through the
National Institutes of Health. Virtually no person, even in the
scientific community, favors complete nonregulation, so I shall
here set that alternative aside. The matter of self-regulation,
however, is an important option, especially when it is considered
that federal regulation of this particular "dangerous" area of
scientific research would in consistency have to be applied to all
similarly dangerous areas of scientific research--whether the
research be in biology, medicine, or the behavioral sciences. The
most important single self-regulatory document in the short history
of the recombinant DNA controversy has been the NIH Guidelines,
first issued on June 23, 1976.[4] Various ways of revising these
guidelines have been proposed and discussed. The most recent and
comprehensive proposed revision was issued in the *Federal Register*
on September 27, 1977 by the National Institutes of Health under
the title "Recombinant DNA Research: Proposed Revised Guidelines."
It was discussed at open hearings on December 15-16, 1977, at NIH.

---

[4]Published in the *Federal Register*, July 7, 1976 (41 FR 27902
et seq.)

In contrast to the sweeping federal controls proposed in Congress, these guidelines cover the following topics:

1. Laboratory containment practices: These include detailed specifications of standard practices used in microbiological laboratories, special procedures and equipment for DNA Research, and various specifications regarding experimental design.

2. Experiments that are not to be performed: *e.g.*, various forms of cloning recombinant DNAs, deliberate release into the environment of recombinant DNAs, etc.

3. Containment guidelines for specific permissible experiments.

4. The roles and responsibilities of the principal investigator and responsible officials in the institution where the research is undertaken.

5. The duties of review groups (study sections) and of the Program Advisory Committee (both NIH).

6. Guidelines on packaging and shipping of recombinant DNA materials.

7. Schemes for the classification of microorganisms on the basis of hazard.

Those who criticize the NIH Guidelines and/or believe in tighter federal controls usually do so because they believe the Guidelines do not go far enough, especially in their containment provisions. Others believe, by contrast, that scientists have been unable to exert reasonable control over the development of these guidelines, having brought up the issue themselves, and now find themselves confronted with an ultra-conservative set of rules that restrains the progress of their research. The topic I shall address in the remainder of this paper is the extent to which federal legislation and tighter controls on DNA research over and above the NIH Guidelines (or some more adequate revision of them)

would be justified. This question of government regulation vs.
self-regulation is a time honored one with many applications
beyond the regulation of scientific research. A central problem
of business ethics, for example, is the extent to which considerations of discretion, equity, and international behavior ought to
be left up to the self-regulatory codes of business adopted by
various business associations or ought to be pursued by federal
regulation. DNA research presents a problem of ethics and science
that parallels this problem of ethics and business, but two important differences should be noted. First, business is usually free
of formal government support, connections, and control (despite
trade associations, federal subsidies, etc.), and its codes of
ethics are detached from federal influence. The National Institutes of Health, by contrast, are parts of the government, though
much committee work surrounding the recombinant DNA guidelines
comes from nongovernment sources. Second, while I would myself on
most occasions favor government regulation of business practices,
I shall offer reasons against government regulation of DNA research.

II

Many believe that strict federal regulations are morally and
prudentially justified in the present circumstance. There seem to
me roughly four reasons why it could be argued that regulation is
justified, though in the end I shall express disagreement with some
of these reasons.

1. *Protection of the Credibility*
    of Scientific Research

Scientific research has not always enjoyed the prestige and
importance, let alone the success it presently does. If it becomes discredited in the eyes of the public, especially in an area

where scientists refused to accept control, the damage to scientific research could be devastating. Also scientific inquiry *can* flourish under fair guidelines partially promulgated by the scientific community itself, but it cannot flourish if it is under constant attack and threat of budgetary cutoffs. Many scientists have themselves asked for federal regulation for this and similar reasons.

2. The Protection of Persons, Animals, and the Environment

It seems reasonably clear that no one has an adequate empirical basis to provide a satisfactory risk/benefit calculus. This being the case, it would seem better to err on the side of caution, especially since the research is not absolutely essential for the achievement of any known major purpose, and certainly should not be entered into with haste. Moreover, though the risks *are* uncertain, *if* there is a substantial or anything approximating a substantial possibility of harm which might result to the general public, then the public's right to limit the freedom of researchers is unquestionable. There also--as previously mentioned--is not and could not be any constitutional protection. No right to inquiry remains a protected right once a clear and present danger has been shown by virtue of an exercise of that right.

There can be little doubt that there is some level of risk that microorganisms with foreign genes could produce disease or cause damage to the environment in the course of DNA research. If they were to escape from the laboratory they could theoretically infect humans, animals, and plants. However, after several years of recombinant DNA research no verifiably dangerous organism has been produced, and even *by analogical inference* from the known effects of known organisms, no level of risk is presently predictable with accuracy. Except for a few dangerous experiments where

there is a known finite probability of harm, estimates of the
dangers of this research have inadequate data bases and are highly
intuitive. Present NIH Guidelines are a policy response to these
scientific unknowns. This second reason, then, can only mean that
we should be sensitive to certain biohazards. It cannot mean that
science should always avoid any level of hazard.

3. *Possible Abuses by Nonexperts*

Most of us do not worry about the continuation of DNA research
in the hands of competent investigators. But the absence of regulation, or even the presence of voluntary compliance, leaves it
open for nonexperts to carry on the research in even primitive
containment conditions. A major objection to the use merely of
the NIH Guidelines is that they are at present enforceable only by
the denial of NIH funding--a provision that seems much too weak.
For example, under this scheme of scientific self-regulation industrial research is not regulated. This state of affairs is inequitable and could become dangerous. Research done on grants is
normally subjected to peer review and to public scrutiny by publication. This protection does not hold for drug and other industrial research, most of which is secret and seldom reported or
scrutinized. Yet it has also been reliably predicted that leading
drug companies expect marketable products from recombinant DNA
research, perhaps within the next five years.[5] These products
will in theory be used to treat human health problems as well as
environmental problems.

[5] Testimony of Dr. Joseph E. Grady of the Upjohn Company, United
States Senate (November 10, 1977).

4. *The Public's Right to Know and*
    Scientific Accountability

Most DNA research is supported by public funds. Since it is
the public's money, we have a right to consent to this use of it.
Just as doctors and research investigators ought to acquire the
informed consent of patients and subjects, so those who perform
recombinant DNA research ought to obtain the informed consent of
the public. The public's interest in research and its public
accountability has not emerged of late in a vacuum. Shortly after
World War II, during the process of the Nuremberg Trials, the
"evils" of careless and unethical research were brought to the
public's attention. Since the development and deployment of
atomic warfare, both scientists and the public have been concerned
not only about the accountability of scientists but even more
importantly about whether, on balance, what scientists produce is
worth it. The recombinant DNA controversy has thus become, as
Stephen Toulmin puts it, "not merely a practical issue, but also a
symbolic one" (1977, p. 102). Historical reasons of this sort
have led to the now firmly entrenched view that informed consent
is a *sine qua non* of justified scientific research involving human
subjects. In the present context the parallel thesis is that the
consent of the public to possibly dangerous experimentation is
absolutely essential. Moreover, just as consent (usually) nega-
tives [negates] injury in the law even when there is harm, so *if*
the public consents it will negative or largely take the sting out
of any claim to have been injured, even if certain harms occur in
the process of performing DNA research. Here scientific accountability begins to merge with the first reason cited above: protection of the image of the scientific enterprise.

Underlying the above four reasons for limiting the liberty of
research scientists is a more philosophical reason which might be
used to justify regulation. The harm principle is an independent
liberty-limiting principle frequently invoked at least implicitly

by the supporters of all sorts of social restrictions on liberty.[6]
It says that coercive interference with persons' liberty is justified if through their actions they would produce harm to other
persons or perhaps to public institutions. It is sometimes argued,
for example, that enactment of this or that program of scientific
research will produce important benefits, while failure to enact
it will produce serious and needless suffering to proximate future
generations or will lead to avoidable and highly expensive costs
to the state. The moral force of such claims is generated by the
harm principle. The harm principle, then, says that when specific
kinds of harm would be caused to a person or a group of persons
the state is justified in intervening, even coercively, for the
purpose of protection. Such harm might be produced in many ways,
but the fact that the harm was produced by negligence or other
inadvertent means would not render interference unjustified.

The role granted to consent is a critical factor in interpret-
ing the harm principle. A person may consent to actions and
still be harmed (as a boxer who consents to a match and winds up
in the hospital is harmed), though we might want to say that he
is not *wrongfully* harmed, because he consented. In my view, the
state's proper role is the prevention either of wrongful harm to
persons or of conditions productive of harm which are unknown or
uncontrollable by the affected persons. The state should not be
in the business of prohibiting or curtailing dangerous activities
when the possibility of harm has been consented to in an informed
manner. I construe the harm principle, then, to allow the state
to intervene only in the former cases and not in cases where
informed consent is present. This is a vital philosophical link
to the final argument (#4) in the previous section: The state
ought not to prohibit or further restrict research, once it has

---

[6]The harm principle and other liberty-limiting principles are
thoughtfully discussed by Joel Feinberg (1973), Chapters 2-3.

been adequately consented to. But it has the duty to control it
prior to consent, in order to protect the public from harm.

III

    Finally, the question must be confronted as to the justifi-
ability of federal regulations now contemplated in addition to the
self-regulatory NIH Guidelines. Although most of the witnesses
at most of the several days of testimony before Senate and House
Committees during 1977 favored the regulation and control of
recombinant DNA research, I do not myself regard federal regulations as needed or justified at this time. While this conviction
could change with any number of advances or accidents in recombinant DNA research, there are three main reasons why the contemplated regulation at federal, state, and local levels does not seem
to me warranted.

    First, recently available scientific information about recombi-
nant DNA research, in this case data pertaining to the *E. coli*
K-12 strain, would indicate that this strain is probably not pathogenic even when genes for known toxins are introduced. The greatest fears about recombinant DNA research have always been that the
research might create an organism of pathogenic capability,
against which known antibodies were inefficacious and which might
produce widespread contagion. At the same time, it has been
thought that the two best maneuvers for avoiding this risk were
proper laboratory containment procedures (hence the major parts of
the NIH Guidelines) and the use of host organisms sufficiently
weakened that they were rendered incapable of survival beyond the
laboratory. Besides pathogenicity studies, other experiments
indicate that the inclusion of foreign DNA decreases rather than
increases the state of fitness of the organism, and it has been
argued that some recombinant patterns already happen in nature
as a common phenomenon. While I am myself incapable of a

scientific assessment of the conclusiveness of these findings, *if* they are basically accurate they would indicate that such organisms are less rather than more likely to survive in most all environments in which they might be placed. It seems likely, then, that experimentation with DNA actually produces less risky organisms, whereas previously it might have been thought to produce more risky ones. This is not an argument to the conclusion that the recombination of DNA molecules presents no risk. Rather, it is an argument that recent research indicates that the risk of using certain strains is lower than was thought, even at the time bills began to flood into Congress.

Second, the wisdom of regulation may be doubted because of its implications for other areas in which legislation might be similarly justified. In moral and legal consistency, all research should receive equal treatment where there is a similar level of possible harm--where "level of harm" is understood in terms of both the probability and the magnitude of the possible harm. The magnitude of possible harm in the case of recombinant DNA seems to many to be great, though the probability may be low. Assuming the principle that when the level of possible risks and benefits is relevantly similar the research should be regulated, much more research should probably be regulated than is now regulated. Research performed in the biomedical and behavioral sciences often entails risk beyond what may be fairly described as minimal; moreover, the possibility of harm is relatively unknown, while the possibility of actually producing benefit is equally uncertain. Many unintended and noncompensable damages can and have occurred in the history of research, even when it has been well designed and executed. If strict regulation had been placed on scientists in the past, many important contributions to knowledge might never have been made, or at least would have been significantly delayed. Special safeguards were erected in the case of research involving radioactive materials, to take one example, and there seems to be no reason why recombinant DNA should be treated differently (by using the guidelines alone as the basis of federal laws).

Third, local institutional review boards have been put in place in the last decade to protect human subjects of research. These committees scrutinize, modify, and (when satisfactory) approve research protocols at local institutions. While the work of these boards has at times been of questionable effectiveness, some have performed well and others are learning rapidly how to scrutinize research. I remain unconvinced that similarly constructed, local biohazards committees are not the best answer to the problem of control over research. Rather than place the problem at the level of federal oversight, local oversights of this sort would probably be both more cost effective and more protective of the public interest. This system provides an ongoing, consistent mechanism for every piece of research involving possible harm to human subjects or the environment. Investigators carrying on research are all treated similarly, whether the research is medical, biological, or behavioral. Local institutions may prohibit the research if it is found too dangerous (or at least may prohibit it from that institution), and research is treated individually on a protocol by protocol basis. I do not see how the risks of research can be closely and fairly inspected, and even closely monitored, unless a local review mechanism is employed. While this proposal is a procedural and not a substantive one (the substantive standards being the NIH Guidelines or their equivalent), it nonetheless seems to me a vital matter of process.

One likely objection to the proposals I have thus far advanced is that they do not bring private industry under surveillance and regulation, as more specific and uniform national legislation might. While true, the import of this objection may be doubted. Federal inspection, licensing, and registration regulations are not likely to uncover that which could now be concealed under the NIH Guidelines. Even if uniform laws existed, those who

would operate outside the boundaries of the NIH Guidelines for
reasons of profit might do so anyway, given the hidden character
of the laboratory setting. [Thus far, the only documented case of
a violation of the Guidelines occurred at a leading academic
research setting (the University of California, San Francisco).][7]
Moreover, should those who engage in such practices cause damage
to persons, animals, or plants in the environment, they would be
subject to prosecution under presently existing laws. Consider
the matter of federal regulation in a broader comparative perspective. Multinational oil corporations might be constantly tempted
by unethical practices, and might thus need the constraint of
federal regulation. By contrast, in recombinant DNA research
investigators are themselves placed at risk by the work (probably
the greatest risk), and there is little direct incentive beyond
curiosity at the present time for uncautious laboratory practices.
Indeed, as Nicholas Wade has pointed out, "even industry . . . is
now, for reasons of self-protection, leaning toward having the
government register and keep track of its gene-splicing activities" (Wade, 1977; 558). Regulations might be more desirable
than at present if the research changes, but until such change
occurs I cannot see even that we ought to err on the side of con-
servatism by the creation of complex legislation.

A simpler and appealing solution to these problems would be to
draft legislation that renders the NIH Guidelines enforceable in
ways other than the mere denial of NIH funds. A streamlined law
could be created that applied to industry, foundations, high
schools, and all nonfunded laboratories. Such legislation would
not have to carry the licensing, monitoring, national commission,
facility-limitation, inspection, and enforcement provisions
presently contemplated by Congress. This course would satisfy
demands of equity and fairness, since, on a national basis, all

[7]See Wade 1977a.

scientists would be treated equally by bringing all research under the Guidelines or their equivalent. Of course there would be other special problems such as how to establish equitable review mechanisms for funded and nonfunded research, and how to award inequitable requirements being imposed by state and local governments (a substantial problem in my view). But presumably these important issues could be resolved short of the highly restrictive federal regulations now contemplated in some quarters. Some would find such legislation unattractive because it would contain no *special* provisions for federal monitoring and enforcing of the NIH Guidelines. But this absence--effectively the current state under NIH Guidelines--is the virtue of this approach, not its vice.

The considerations I have advanced reduce largely to the following: our society quite appropriately protects scientific research through its commitment to freedom of inquiry. Those who would curb or prohibit such inquiry at the federal level are obliged to show that real and not merely imagined dangers are present in such research and that the hazards are of such moment that they cannot be (1) locally controlled and (2) regulated by rules promulgated by scientists themselves. The recombinant DNA controversy is one example of the general problem of the public accountability of scientists, but similarly it is an example of the burden of proof being on those who would place special liberty-limiting restriction on research. That burden of proof does not seem to me to have been met.

REFERENCES

Feinberg, Joel. 1973. *Social Philosophy*. Englewood Cliffs, New Jersey: Prentice Hall.

"Lobbying Derails Bills on DNA Regulation" (SGR, 1977). *Science and Government Report, 7* (October 15), 1-2.

Toulmin, Stephen. 1977. "The Research and the Public Interest."
   In *Research with Recombinant DNA*. Washington, D.C.: National
   Academy of Sciences.

Wade, Nicholas. 1977. "Gene-Splicing: At Grass-Roots Level A
   Hundred Flowers Bloom." *Science, 195* (February), 558-560.

_____. 1977a. "Recombinant DNA: NIH Rules Broken in Insulin
   Gene Project." *Science, 197* (September), 1342-1345.

_____. 1977b. "Confusion Breaks Out Over Gene Splice Law."
   Science, 198 (October), 176.

Regulation at Cambridge

David Clem

City Council
Cambridge, Massachusetts

I have been asked to speak on the problems in regulation of recombinant DNA research at the local level, specifically in Cambridge, Massachusetts where I am a member of the Cambridge City Council and Chairman of the Ordinance Committee that developed controls for recombinant DNA research. I am not a scientist nor a philosopher; I think I am a pretty good politician.

However, I am reminded of another person named Clem--Bill Clem--the professional baseball umpire, who is said to have remarked, "There are strikes and there are balls but they ain't nothing until I call them." In November of this year the voters of Cambridge will tell me if I have struck out on the issue of recombinant DNA research.

In the middle of the controversy in Cambridge over the continuation of certain types of experiments utilizing the techniques of recombinant DNA, I received a telegram from someone in Texas that read:

241

Stop Recombinant DNA Research. There are already more
forms of life in Harvard Square than this country can stand.

I can assure you it was often tempting to resort to simplistic
solutions such as this, but the Cambridge City Council somehow
persevered and created a model for participatory decision-making
that has received broad acclaim. In many ways the events in
Cambridge precipitated a wide, public debate about recombinant
DNA research--a debate that brings us to this seminar.

Therefore, I would first like to outline what actually
happened in Cambridge so that you might better understand the
utility of the Cambridge experience. It is my contention that an
important lesson to be derived from the Cambridge experience has
been largely overlooked in this seminar.

What is the lesson to be learned from Cambridge? It is
simple, straightforward, and, I believe, quite sound. The lay
public has a right to be involved in decisions affecting their
general health and safety and, given adequate information and
meaningful power, the lay public can address complex issues and
resolve them equitably. The proof is in the pudding.

In June of 1976 the Cambridge City Council became aware of an
intense, and sometimes acrimonious debate within the Harvard
scientific community. The issue was the proposed construction of
a P-3 laboratory at Harvard to facilitate recombinant DNA
research. I think it is fair to say that in spite of the wide
coverage within the scientific journals of the NIH process to
articulate research guidelines, the overwhelming majority of the
Cambridge City Council had never heard of recombinant DNA
research. None of us understood the profound implications of
conducting such research. We asked to be informed, not a radical
demand, and the Mayor convened two public hearings on the issue.

What the Council learned was genuinely disturbing. First, we
discovered that several prominent scientists were opposed to
allowing recombinant DNA research to continue. The original call

for limits to inquiry came, not from the Cambridge City Council, but from the academicians themselves.

Second, we learned that the issue was not new. For over two years the scientific community had debated the safety aspects of continuing recombinant DNA research and had even imposed a voluntary moratorium on certain types of experiments. Therefore, the concept of a moratorium came not from the Cambridge City Council but from the scientists themselves.

Third, the City Council learned that the NIH had promulgated guidelines for conducting recombinant DNA research. However, we soon discovered that these guidelines only applied to NIH funded projects and were not applicable to industrial research, private nonprofit research, or research funded by other federal agencies. Assuming guidelines were warranted, the NIH Guidelines, released the day of our first public hearing, were not comprehensive. Indeed, the first call for broader acceptance of the guidelines came not from the City Council but from the director of the NIH.

Fourth, the City Council was informed that recombinant DNA experiments at the P1 and P2 levels were already underway at Harvard and at MIT and, in spite of the requirements of the Environmental Protection Act of 1970, the director of the NIH, had released the guidelines prior to the completion of an environmental impact statement. The City Council did not charge the NIH of violation of public policy, if not law; the director of the NIH *admitted* violation because of compelling health issues and the need for uniform safety standards.

Fifth, the Cambridge City Council gradually understood that the NIH Guidelines were basically developed by scientists performing recombinant DNA research or directly interested in recombinant DNA research. We discovered that monitoring of compliance to the Guidelines was left to the institutions performing the work and that the only penalty for noncompliance was removal of funding, executed through the "courts" of the NIH. The researchers were their own judge and jury. The City Council did not conjure up a

violation of the principle of self-interest; it was self-evident
and self-imposed.

So what did the Cambridge City Council do?

Did we ban all recombinant DNA research? No.

Did we stop any scientist from continuing research already
underway? No.

Did we prohibit Harvard from building their laboratory? No.

Did we force MIT to withdraw its petition to the NIH for
laboratory certification? No.

Did we ask for a good faith moratorium on P3 and P4 research--
research *contemplated* but not underway? Yes.

Did we allow Harvard to begin construction on their laboratory,
subject to regular building and health inspections to insure
normal compliance to state building laws and the recently released
NIH Guidelines? Yes.

Did we create a Cambridge Experimentation Review Board, con-
sisting of lay citizens completely removed from recombinant DNA
research, to review the NIH Guidelines and to report back to the
Council on recommendations for action? Yes.

Did we divorce this study group from opportunistic political
interference and give them complete latitude in tackling this
difficult and confusing, and complex issue? Yes.

That's what the City Council did--a housewife, a student, a
security guard, a pharmacist, a college administrator, a welfare
mother, a rubbish disposal contractor, a court clerk, and a tax
collector--nine city councillors in Cambridge, Massachusetts.

Now before someone argues that Cambridge is somehow a special
case because of its unique academic environment--the City of

Harvard and MIT--let me educate you about the Cambridge most out-
siders do not know.

Cambridge is a city of 100,000 people, 30,000 of whom are
students or university related. One-third of our population
works in blue collar jobs. Fifteen percent of all Cambridge resi-
dents live in public or subsidized housing. The median family
income is $9,865/year. Of the 100 largest cities and towns in
Massachusetts, Cambridge ranks in the top five for every single
measurement of poverty. Our population is 9.7% Black; 5.7% is
Hispanic; 5.4% is Portuguese; 3.2% is Greek. We have five bi-
lingual programs in our public schools. The City of Cambridge
has been described not as a melting pot but as a tossed salad with
very distinct ingredients blended together, but still maintaining
their unique characteristics.

The Cambridge City Council created a citizen review group and
called it the Cambridge Experimentation Review Board. It consisted of lay citizens who represented a slice of life of the
Cambridge community with no involvement in recombinant DNA
research or with the institutions proposing to do the research.
We gave them six months to do their work.

Some of us on the Council were initially worried about passing
the buck to a review committee. I am mindful of the definition of
a committee as a group of people who individually can do nothing,
but acting as a group decide collectively that nothing can be
done. We, as members of the City Council, were pleasantly surprised and, not surprisingly, reminded of our own roots.

For almost six months, eight Cambridge citizens volunteered
two nights a week to learn, to question, to debate, and to
decide. Their recommendations were *unanimous*.

Let me quote from the introduction of the CERB Report:

While we should not fear to increase our knowledge of the
world, to learn more of the miracle of life, we citizens
must insist that in the pursuit of knowledge appropriate

safeguards be observed by institutions undertaking the
research. Knowledge, whether for its own sake or for
its potential benefits to humankind, cannot serve as a
justification for introducing risks to the public
unless an informed citizenry is willing to accept those
risks.

Was the group satisfied only with a risk-free environment? No.
And again I quote from the report:

We recognized that absolute assurance was an impossible
expectation. It was clearly a question of how much
assurance was satisfactory to the deliberating body, and
in the case of CERB, that body was comprised of citizens
with no special interests in promoting the research.

The CERB recommended:

1.   That *all* recombinant DNA research conducted in the City of
Cambridge be declared a health hazard unless it was conducted
under strict compliance to the NIH Guidelines. (This closed the
loophole of non-NIH funded research and shifted the burden of
proof to the researchers to show cause that the research was safe.)

2.   That additional guidelines be imposed in Cambridge.

   a.   Mandatory training in procedures for minimizing potential accidents for *all* laboratory personnel and the preparation of a training manual which contains the NIH-approved
procedures for recombinant DNA research.

   b.   Expansion of the institutional biohazards committee
mandated by NIH to include members from a variety of disciplines, representation from the bio-technicians staff, and at

least one community representative unaffiliated with the institution and selected by the Health Policy Board of the City of Cambridge.

    c. An EK2 level of biological containment for all experiments undertaken at the P3 level.

    d. Ongoing screening to insure the priority of the strain of host organisms and testing for the resistance of experimentally created organisms to commonly used therapeutic antibiotics.

    e. A health monitoring system to monitor the survival and escape of the host organism in the laboratory worker including available means to monitor the intestinal flora of the laboratory worker.

3. That a Cambridge Biohazards Committee (CBC) be established for the purpose of overseeing all recombinant DNA research that is conducted in the City of Cambridge.

    a. The CBC shall be composed of the Commissioner of Public Health, the Chairman of the Health Policy Board and a minimum of three members to be appointed by the City Manager.

    b. Specific responsibilities of the CBC shall include:

    i. Maintaining a relationship with the institutional biohazards committees.

    ii. Reviewing all proposals for recombinant DNA research to be conducted in the City of Cambridge for compliance with the current NIH Guidelines.

iii. Developing a procedure for members of institutions where the research is carried on to report to the CBC violations either in technique or established policy.

iv. Reviewing reports and recommendations from local institutional biohazards committees.

v. Carrying out site visits to institutional facilities.

vi. Modifying these recommendations to reflect future developments in federal guidelines.

4. That the City Council of Cambridge, on behalf of this Board and the citizenry of the country, make the following recommendations to the Congress:

a. That all uses of recombinant DNA molecule technology fall under uniform federal guidelines and that legislation be enacted in Congress to insure conformity to such guidelines in all sectors, both profit and nonprofit, whether such legislation takes a form of licensing or regulation, and that Congress appropriate sufficient funding to adequately enforce compliance with the legislation.

b. That the NIH or other agencies funding recombinant DNA research require institutions to include a health monitoring program as part of their funding proposal and that monies be provided to carry out the monitoring.

c. That a federal registry be established of all workers participating in recombinant DNA research for the purpose of long-term epidemiological studies.

   d. That federal initiative be taken to sponsor and fund
research to determine the survival and escape of the host
organism in the human intestine under laboratory conditions.

The Cambridge City Council, in spite of strong personal disagreements with some of the conclusions of the CERB report on
both sides of the issue, *unanimously* adopted a local ordinance
implementing the report.

What does this series of actions suggest? Again I turn to the
CERB report:

A predominantly lay citizen group can face a technical scientific matter of general and deep public concern, educate itself
appropriately to the task, and reach a fair decision.

Let me conclude by trying to articulate some principles or
criteria that guided the Cambridge City Council.

   1. The Principle of Informed Consent--Right to Know. This is
fundamental to our democratic process. Recent revelations about
our CIA, our national leaders, Department of Defense experiments
no doubt heightened our determination to fulfill this right.

   *2. The Inevitable Predicament of Self-Interest Decision-
Making--The Need for Accountability.* Scientists are no different
from other segments of our society. We do not allow automobile
manufacturers to dictate pollution control standards or the
generals to be commander-in-chief. We should not expect scientists to be immune from this phenomenon.

   *3. Malicious Intent Is Not the Only Criterion for Imposing
Regulations.* Toy manufacturers do not deliberately create toys
that cut children, put out eyes, or explode into flames. Prudence
suggests that mistakes, abuses, and unforeseen consequences will

happen and that outside observers are more vigorous in controlling for these negative consequences.

4. *Separation of Powers*--an agency that promulgates research cannot be expected to regulate research. Events surrounding the AEC should have taught us this principle.

Beyond Recombinant DNA

Straddling the Boundaries of Theory and Practice: Recombinant DNA Research as a Case of Action in the Process of Inquiry[1]

Hans Jonas

**The New School for Social Research
New York, New York**

Recombinant DNA research opens a new chapter in the long story
of the fusion of theory with practice that has been going on in
the progress of modern science. It thus poses anew the question
of the "freedom of inquiry." That freedom is cherished by the
Western world as part of its general regard for freedom. Every
freedom, to be sure, is in its exercise qualified by responsibility
and social restraints. But the freedom of inquiry seemed to be
exempt from such qualifications on the premise that, in purpose
as well as in procedure, it is clearly divorced from action and,
thus, can hurt nobody. Truth by itself, useful or not, is a
supreme right, even a duty, and ideally the only right that will
not encroach on the rights of others. Hence, freedom within this
enclave can be total. Let us take a critical look at the premise,
bearing in mind that "inquiry" today means preeminently *scientific*
inquiry in the technical sense.

[1]To appear: G. Winkel, D. Glass and A. Fulco (Eds.), *Social and
Ethical Implications of Science and Technology* (New York: Plenum).

I. THE FUSION OF THEORY AND PRACTICE IN MODERN SCIENCE

The traditional argument was that science does not intersect
with morals in general, beyond the internal morality of keeping
faith with the standards of science itself. Its sole value is
knowledge, its sole business the pursuit of it. This does indeed
impose its own code of conduct which can be called the territorial
morals of the scientific realm: abiding by the rules of method and
evidence, perservering, not cheating, etcetera--in sum: being
rigorous and intellectually honest. These virtues are conditions
of good science and imply no commitment beyond it. The imperative
is simply: when a scientist, be a scientist. So considered,
science constitutes a moral island by itself.

But this self-image of science is not the whole truth. Something like it was true so long as the contemplative sphere and the
active sphere were cleanly separate (as they were in pre-modern
times), and pure theory did not intervene in the practical affairs
of men. Knowledge could then be considered a good private to the
knower, which could do no harm to the good of others. Its business was to comprehend, not to change, things. Knowledge and the
acquisition of it, by observation and reasoning, were states of
mind, as such communicable, but not interventions in the state of
things. However, the rise of natural science at the beginning of
the modern age changed this traditional relation of theory and
practice, making them ever more intimately merge. We still pay
homage to the dignity of "knowledge for its own sake." But it
would be hypocritical to deny that in fact the emphasis in the
case for science has heavily shifted to its practical benefits.
From the Industrial Revolution onward, there was an increasingly
irresistible spill-over from theory, however pure, into the vulgar
field of practice in the shape of scientific technology. Belated-
ly and almost suddenly, Francis Bacon's (1561-1626) precocious
directive to science to aim at power over nature for the sake of
raising man's material estate had become working truth beyond all

expectation. Therewith, the subject of "science and morals"
begins in earnest. For whatever of human doing impinges on the
external world and thus on the welfare of others is subject to
moral and legal rules. The very praise of the benefits of science
exposes science to the question of whether *all* of its works are
beneficial. It is then no longer a question of good or bad
science, but of good or ill effects of science (and only "good
science" can be effectual at all). An often-heard question is:
If technology, the offspring, has its dark sides, is science, the
progenitor, responsible for them?

The simplistic answer is that the scientist, having no control
over the application of his theoretical findings, is not responsible for their misuse. His product is knowledge and nothing
else: its use-potential is there for others to take or leave, to
exploit for good or evil, for serious or frivolous ends. Science
itself is innocent and somehow beyond good and evil. Plausible,
but too easy. The soulsearching of atomic scientists after
Hiroshima tells as much. We must take a closer look at the inter-
locking of theory and practice in the actual way science is nowadays "done" and essentially must be done. We shall then see that
not only have the boundaries between theory and practice become
blurred, but that the two are now fused in the very heart of
science itself, so that the ancient alibi of pure theory and with
it the moral immunity it provided no longer hold.

The first observation is that no branch of science is left
whose discoveries have not in them some technical applicability.
(The only exception I can think of is cosmology.) Every unravel-
ling of nature by science now invites some translation of itself
into some technological possibility or other, often even starts
off a whole technology not conceived of before. If this were all,
the theoretician might still defend his sanctuary this side of the
step into action: "That threshold is crossed after my work is
done and, as far as I am concerned, could as well be left
uncrossed." But he would be wrong. What is the true relationship?

First, much of science now lives on the intellectual feedback from precisely its technological application. Second, it receives from there its assignments: in what direction to search, what problems to solve. Third, for solving them, and generally for its own advance, it uses advanced technology itself: its physical tools become ever more demanding. In this sense, even purest science has now a stake in technology, as technology has in science. Fourth, the cost of those physical tools and their staffing must be underwritten from outside: the mere economics of the case calls in the public purse or other sponsorship; and this, in funding the scientist's project (even with "no strings attached"), naturally does so in the expectation of some future return in the practical sphere. There is mutual understanding on this. With nothing shamefaced about it, the anticipated pay-off is put forward as the recommending rationale in seeking grants or is outright specified as the purpose in offering them. In sum, it has come to it that science has its tasks increasingly set by extraneous interests rather than its own internal logic or the free curiosity of the investigator. This is not to disparage those extraneous interests nor the fact that science has become their servant, i.e., part of the social enterprise. But it is to say that the acceptance of this functional role (without which there would be no science of the advanced type we have, but also not the type of society living by its fruits) has destroyed the alibi of pure, disinterested theory and put science squarely in the realm of social action where every agent is accountable for his deeds.

Even that is not all. The involvement of scientific discovery with action goes deeper than via its eventual application. How does the scientist get his knowledge? There was a time when the seekers after truth had no need to dirty their hands, of which noble breed the mathematician is the sole survivor. Modern natural science arose with the decision to wrest knowledge from nature by actively operating on it, i.e., by intervening in the objects of knowledge. The name for this intervention is "experiment," vital to all modern science. Observation here involves manipulation. Now, the grant of freedom to thought and speech, from which that of inquiry derives, does not cover *action*: that always was and remains subject to legal and moral restraints. Originally, experimentation kept to inanimate matter, and there to small-scale, vicarious models contained in the laboratory, which still secured some insulation of the cognitive arena from the real world. But experiments nowadays can be less innocuous and indeed ambiguous as to their being mere experiments. An atomic explosion, be it merely done for the sake of theory, affects the whole atmosphere and possibly many lives now or later. The world itself has become the laboratory. One finds out by doing in earnest what, having found out, one might wish not to have done. And as to animate objects, experimentation inevitably deals with the original, not with surrogates, and ethical neutrality ceases at the latest when it comes to human subjects. What is done to them is a real deed for whose morality the interest of knowledge is no blanket warrant. In short, the very means of getting to know may raise moral questions even before the question how to use the knowledge poses itself.

From both ends, therefore—that of its technological fruits and that of its own techniques of preparing the theoretical soil for them—modern science finds itself exposed to the winds of ethical challenge.

II. MORAL ISSUES IN THE CONDUCT OF INQUIRY

So far we have seen how, in modern science, man's quest for knowledge lost its time-honored purity and became thoroughly alloyed with mundane action. Not only in what it seeks knowledge about—already in how it obtains it, the line between thought and deed often vanishes. This must affect the venerable "freedom of

inquiry." We are chary of interfering with it and should indeed
do so as little as possible--both for the sake of science, which
can only thrive on autonomy, and for the sake of mankind, whose
cause in more senses than the utilitarian is wedded to the
advancement of knowledge. Yet we must remember that complete
immunity of theory depends on its seclusion from practice. Never
has absolute freedom been claimed for action, and surely never
been accorded to it. Thus to the extent that science becomes
shot through with action, it comes under the same rule of law and
social censure as every outward acting in civil society. Obviously, this bears on the admissibility of experiments, which are
not necessarily innocent because they promote knowledge.

To make the point just by citing notorious atrocities is to
weaken it. One easily agrees, *e.g.*, that one must not, in order
to find out how people behave under torture (which may be of
interest to a theory of man) try out torture on a subject; or kill
in order to determine the limit of tolerance to a poison, and the
like. Remembering Nazi research in concentration camps, we know
too well that the perpetrators of such scientific experiments were
despicable and their motives base, and we can wash our hands of
them. Here was "freedom" of inquiry as shameful as its worst
suppression. Arguably even, the case falls outside science and
wholly into human depravity. Our problem is not with that, nor
with crooked or perverted science, but with bona fide, regular
science. Keeping to indubitably legitimate and even praiseworthy goals, we ask whether in *their* pursuit one may, *e.g.*,
inject cancer cells into noncancerous subjects, or (for control
purposes) withhold treatment from syphilitic patients--both actual
occurrences in this country and both possibly helpful to a
desirable end. Whatever the answer, it is obvious that here
moral and legal issues arise in the inner workings of science--
issues that crash through its territorial barriers and present
themselves before the general court of ethics and law.

Biomedical research is the most fertile ground for these kinds
of issues. Medicine, of course, is by definition not a disinterested science but committed to a goal sanctioned by every
standard of private and public good. But it relies heavily on
scientific research which, although geared to those practical
ends, has its component of pure theory. In that respect, medicine
is a branch of biology. This in turn, once mostly a theoretical
discipline, is becoming increasingly pregnant with utility potentials. Applied biological knowledge, medical or otherwise, is a
technology--to which, as we have found, theoretical inquiry is
then wedded. What better use can there be for a science than to
benefit its very subject when this is life itself? Yet, no
scientific-technological alliance is so rife with moral problems
(blatant abuses discounted) as that of the life sciences, from the
conduct of research all the way down to the last decisions on
uses. It begins even prior to research with the allocation of
finite resources: priorities must be settled among competing projects. The decisions are societal, not intrascientific, and
cannot fail to be morally weighted. A crash program in cancer
research? Or a general improvement in health services? Here both
goals are in themselves flawless (though the all-out search for a
cancer cure may serve to deflect from the less glamorous attack on
man-made cancer causes in the environment, which would clash with
powerful economic interests). There are also disputable goals.
Prevention and interruption of pregnancy are not, by the original
meaning of medicine, properly medical goals, unless pregnancy be
equated with disease and the fetus with a tumor. They may be
approved nonetheless, on nonmedical grounds. Pursuing research
toward them implies a tacit option for birth control, free sex,
free abortion--surely choices in ethics and social policy.

The choice having been made, research itself poses its ethical
problems with its need to experiment on human subjects, present
and future. And here a point can be reached where a research goal

becomes inadmissible merely because it requires inadmissible
experiments.

A case in point is genetic research when it seeks to determine, *e.g.*, whether human cloning is possible, or improvement of
the human type by "genetic surgery," *i.e.*, modifying the gene composition in reproductive cells. At least *one* try at real cloning
or at really producing a genetically altered individual is necessary to find out: the very deed eventually to be decided on in the
light of knowledge is already committed in the night of ignorance
for obtaining the knowledge; and the first clone or genetic freak,
experimentally produced, is as real and definitive as any individual
brought forth into the world. Even discounting the overwhelming
risk of beginning with monstrosities before the technique is per-
fected (without the moral freedom of hardware engineering to
scrap the failures), there is simply no right to experiment on the
unborn--nonconsenting by definition--and for this reason alone
this whole venture is ethically unsound. We pass over the more
philosophical objections against the goals as such.

Returning from these extravagant, futuristic perspectives to
present realities, we have the problem of consent which besets
even the most defensible experiments on human subjects and is
bound up with the mechanics of recruiting them. The law pre-
scribes "informed consent." But who can be really "informed,"
i.e., fully understand, except fellow scientists, who should
indeed be the first to volunteer? But in mere point of numbers,
this recruiting base is too small statistically. Next best are
the educated classes--"professionals" mostly, who also are
socially best placed to satisfy the second ethical requirement,
viz., that the consent be "voluntary." But for obvious reasons,
numerical and other, actual recruiting falls back on more captive
populations: students, welfare patients, prison inmates, where
freedom of consent (=freedom to refuse) is questionable and for
the last two categories the meaning of "informed" is often almost

empty. Here lies a twilight zone of great ethical vulnerability
for much of today's vital research.

Often the research goal itself falls into the twilight zone.
Behavior control is such a goal. It may be socially useful and
easily too useful, *e.g.*, for saving on good government by engineered docility. But even apart from such abuses (not abuses by
the lights of Skinner), the whole *concept* of behavior control is
in tension with such ultimate values as personal autonomy and
dignity. It is, therefore, quite in order to *ask* whether scientific inquiry should move in that direction at all: again, a
question of ethics outside the jurisdiction of science.

One research goal with powerful appeal but ethical pitfalls
concerns aging and dying. Averting premature death is a prime
duty of medicine. But to latest biological thinking there is
nothing definite about a "natural" span of life, and to its
theoretical hope for control of aging every death is "premature."
Leaving undecided whether indefinite longevity is an unalloyed
good for the individual, we look at the social price that finite
living space will exact: proportionate diminishing of births,
hence of youth and new beginnings in the aging social body. Is
that good for the human cause? Whatever the answer, it should
influence the goal choices of scientific inquiry. Here and elsewhere (not confined to the life sciences) we come up against the
moral interface between science and society.

III. THE CASE OF RECOMBINANT DNA RESEARCH

"Recombinant DNA," the latest arrival on the scene of basic
research, fits into the general picture we have drawn and at the
same time adds a new twist to it. The general picture, to restate
it once more, was that in the activities of modern science the
line between thought and deed no longer runs between the concluded
business of inquiry ("pure science") and the subsequent

utilization of its findings in mundane technology, but is crossed
routinely in gaining the knowledge in the first place. The line,
in effect, has vanished inside the cognitive endeavor itself, and
succeeding application follows almost without a break from the
nature of the knowledge so obtained. With this *intrinsic* "technological" quality of the epistemic process, it is of secondary,
mostly subjective importance whether the investigator's own motivation was more "theoretical" or more "practical." Its practical
afterlife is anyway prefigured, even pre-enacted, in the theoretical phase. His was a scouting for the practitioners to follow.
The actual step, though, into the common world of use is still
subject to human agency and its decisions on the inert objects
of the knowledge acquired. This is not so sure to hold for recombinant DNA research anymore, and that is a main cause of the public uneasiness: Here, the experiment may of itself become definitive reality, and at that, reality of a novel and unpredictable
sort. This peculiarity sharpens the "fusion of theory and practice" which the new research shares with most of modern science.
We think of these points:

1. The *aim* is preponderantly practical: to evolve a capacity
for *making* something hoped to be useful (for health, agriculture,
etc.), and the gain to theory ensues as a by-product of practical
success.

2. The *method, i.e.*, the way to knowledge, is the actual
making of the very entities about which knowledge is sought and
whose usefulness is then to be tested.

3. The *entities* so produced inside the research context are
not inert, waiting for us to put them to use, but being alive are
active on their own, so that potentially they may effect their
entry into "practice," *i.e.*, into the outside world, by themselves
and take the decision about use or non-use out of our hands.

4. In the theoretically not excluded eventuality of gene
splicing on *human* gametes or zygotes, if allowed to grow to their
terminal phenotypes, the resulting "Chimeras" (not just "Chimeras
DNA") would even in the first, experimental instance constitute
ultimate deeds that entirely transcend noncommittal theory.

Leaving aside the last point with its dizzying perspectives, let
us enlarge on the first three in their strictly microbiotic scope.

1. Speaking of a preponderantly practical aim is not to deny
a genuine theoretical interest in recombinant DNA research.
There is every reason to expect added insights into the innermost
mechanisms of life from this kind of manipulative investigation,
as we can generally expect from a controlled changing of selected
variables for any complex natural phenomenon. But in the debate
on hazards, one finds major emphasis being given to potential
benefits as the justification for going ahead with this line of
research, even as a moral argument against slowing its pace. As
to the invoked interests of pure *theory*, the question can be asked
whether its proper aim, viz., to understand what life *is*, cannot
be served in the more conservative (if less speedy) manner of
working with its given patterns instead of the revolutionary
manner of creating new ones. Scientists insist that the latter,
innovative road is decisive at this stage for progress in basic
theory, and the layman cannot quarrel with them. Anyway, and
under whichever flag, work on recombinant DNA is here to stay, and
man-made *novelty* is the meaning of "recombinant": to synthesize
new organisms. Even if only disinterested curiosity were invoked
for having and observing them, it must then at least be noted that
the meaning of "theory" has been enlarged, from understanding what
is, to trying out what *could* be: surely a less self-evident and
more gratuitous goal of man's cognitive quest.[2] But in fact, as

[2]Unravelling the DNA structure, a *datum* of nature, was a feat

few will doubt, the dominant thrill lies in finding out what these
novel things can *do*, what consequently *we* could do with them—that
is, in their preconceived practical promise. This promise indeed
specifies their planned design to begin with, *e.g.*, what gene from
one species to select for transplanting into the genetic machinery
of another: a feat of effect-oriented engineering rather than free
theoretical inquiry. And it is the dazzling promise—of the
bacterial hormone factory, of the nitrogen fixing bacterium and
its suitably redesigned host plant—that is argued against the
risks.

   2. This leads to the second point. For finding out what
these things can do one must first make them exist, even show that
they *can* exist, demonstrating their possibility by the accomplished
fact. The theoretical investigator thus turns practical creator
in the act of investigation itself. No simulated models, no
vicarious representations will do in this area, only the true
entities themselves in the fullness of their capacity, which they
will show when set to work. Thus the experiment, different from
its proxy role in conventional research, coincides here with the
very generation of the objects to be investigated; the cognitive
operation is an originative *making*. This is itself a novum in the
history of knowledge. All science, it is true, involves, beyond
intellection, external action inasmuch as there is experimentation
in it, and in that sense it has long since passed outside the contemplative bounds. But the present case represents the further
step that the action inside investigation produces in earnest the
very reality which the normal experiment takes as a given.

of theory in the classical sense: gene splicing is an artificer's
technique first, with the artefact to become an incremental *datum*
in nature's inventory.

3. Add to this the third point: that unlike other artifacts
the reality thus created, a new entry in the general chart of
existence, is *alive,* that is to say, self-active, self-multiplying,
and spontaneously interacting with other life--and we see that
here the element of action in research has a thrust of its own
beyond the research situation and that its laboratory inception
is pregnant with indefinite propagation in the real world. Not
just a new entity--a new agency is being inserted in the balance
of things: tentatively at first in the confinement of the lab, but
once discharged, by accident or design, then in full and perhaps
irrevocable earnest.

Clearly then, and without yet considering the awesome proposition of human-genetic engineering, recombinant DNA research on
microorganisms alone does fit with a vengeance our previous
description of how science has forsaken the sanctuary of pure
thought and become enmeshed with action that is subject to extra-
scientific criteria of the public good. This itself is uncon-
tested, but less so how this impinges on the cherished conception
of total intrascientific freedom of inquiry. The scientific community, anxious about the autonomy of research as a whole, favors
self-legislation from its own ranks and can claim that it has
shown itself aware of its public responsibility with no outside
pressure. Indeed, scientists themselves--those most directly
engaged in the work--were the first to raise the issue and advocate before their peers the need for safeguards, controls, and
even prohibitions, which are invested with an authority beyond
the individual scientist's conscience. I take the story to be
well-known, from the memorable call for a voluntary moratorium by
a committee of the National Academy of Sciences in 1974, the
Asilomar conference in 1975, the promulgation of "Recombinant DNA
Guidelines" by the National Institutes of Health, and the broaden-
ing public debate since then. Too much of the public interest was
involved to keep the matter within the family. On the part of

what one vaguely calls the scientific community, by no means a
homogeneous body, the prevailing consensus seems to be this:

1.   The *risk-benefit* computation strongly comes out in favor
of the research to go on as the mere scientific fascination urges
it to do.

2.   *Safeguards* against the hazards (if any exist different in
degree or kind from what other research has had to guard against)
can be made sufficiently effective by way of physical and biological containment. They can specify "super-safe" facilities as
mandatory for experiments with high-risk viruses. And they can
ban outright use of the most dangerous organisms.

3.   Defining the safety *rules* and *enforcing* them should not be
by law and public authority but by agreement and peer review within the scientific establishment.

I am not qualified to judge the first two points, the risk-benefit
balance with its great expectations on the benefit side (dissent-
ing voices are heard from the scientific camp itself), and the
effectiveness of the safeguards when scrupulously observed (which,
of course, brings in the less calculable human element). And it
stands to reason that any rules to be drawn up will have to rest
on scientific expertise in the first place (with some vigilance
for the bias of group interest). But on the enforcement angle I
can venture a citizen's opinion. It is this.

The stakes being what they are, the public interest clearly
impels public control of this hazardous field (granted even that
up to now, only possible and not demonstrably existing hazards
can be spoken of), and the only coercive instrument of control,
imperfect though it be, is the law. The "scientific community,"
in spite of its hitherto blameless credentials, is not the kind of
body that can vouch for its members or wield effective sanctions

against trespassers. It is a community by virtue of professional
intercommunication only, but otherwise is the mere sum of practicing scientists everywhere and in no way institutionalized (not
even nominally unified by a code of ethics or something like the
Hippocratic oath). Guidelines, such as issued by the National
Institutes of Health, can only count on voluntary compliance,
except where the strong arm of funding gives them muscle. But to
rely on voluntary compliance on the part of great and scattered
numbers of headstrong individuals, answerable only to themselves,
is unsound. Not counting simple sloppiness, the temptations to
recklessness are great. The race for breakthroughs is on, Nobel
prizes beckon, not to speak of material rewards,[3] and he who cuts
corners may come in first. Academic institutions can probably be
trusted with conscientious self-policing. The pharmaceutical
industry surely less so; peer review does not reach there and
would anyhow carry little weight. And then there are the non-
affiliated, lone researchers. For this, it should be noted, can
be a basement operation, in physical size and cost. No mighty
apparatus is needed here, as it is in nuclear physics, with its
mile-long accelerators, numerous staff, public funding--all this
insuring fullest visibility and control. But to handle restriction enzymes and to splice genes, I have been told, is really quite
simple, almost any biological lab and any skilled microbiologist
can do it, and do it clandestinely.

Here, then, I submit, is an obvious case for public regulations
with teeth in them. Offhand one might think of a statutory
licensing system, for which there are many precedents in our
society. The practice of medicine and of law, even driving a car,

[3]There would be, said one scientist at the Asilomar conference,
showing the nitrogen-fixing bacteria nodules on the roots of a
legume plant, "a million dollar prize for anyone who put that kind
of a nodule on a corn plant, eliminating the need for chemical
fertilizers" (*The Washington Post*, March 9, 1975, p. B2).

require a license for which definite standards must be shown to be
satisfied, and doing these things without a license is a criminal
offense. And not only the driver, the car also must pass inspection. I see no reason why something similar should not apply to
facilities and personnel in recombinant DNA research. Only certified researchers in certified labs would be permitted by law to do
this research, with imposed rules of procedure whose observance is
open to verification, and whose violation is actionable. The misgivings of scientists about such outside intervention weigh little
against the misgivings of the public about the lurking and unknown
hazards. Responsible scientists should actually welcome a distinction between legitimate and illegitimate research, the out-
lawing and penalizing of the latter, and should bear the nuisance
of its administrative aspect with good grace. Almost certainly,
in the nature of the case, the instrument of supervision would
still be "peer review," *i.e.*, inspection and adjudication by fellow
scientists, but under government authority and backed up by the
judicial powers of the law.

Would any legislation and the apparatus it sets up be effective? Admittedly, no law prevents all transgression, but defining
transgression and making it punishable do make a difference--and
would do so even given the elusiveness to public notice we have
claimed for this particular type of research when arguing against
trust in self-policing by the imaginary body scientific. If that
argument was correct, then by the same token individuals would
not find it difficult to break the law and escape detection.
Still, the voice of the one and the threat of the other can tilt
the balance, especially where a whole profession needs for its
life a climate of societal approval, thus a record of law-
abidingness. (No spy system is advocated, though reporting violations of the law may become a citizen's duty in the public
interest.) In any case, in an imperfect world, even too wide a
net is better than none.

Admittedly also, the implication of a regulatory system, its
effectiveness apart, is bound to be cumbersome, as is the way with
bureaucracies. I was asked at a legislative hearing: will this not
cause delays in the scientific process and perhaps postpone
results of high value to the public good? My answer was: Almost
certainly the first and quite possibly the other, but that is a
tolerable price for safety. We, if not the impatient scientist,
can wait a little longer for those results, those new organisms,
without which we have managed for so long. "I don't regard it as
sinful to postpone benefits," said Harold P. Green (professor of
law) at the Asilomar conference-- to the displeasure of many
scientists, it seems. I agree with him. Progress, it must be
remembered, is an optional value, certainly its tempo in this
case: we are not in a rush. The situation is not, after all, as
it was with the atom bomb in World War II. There, hurry was of
the essence, to beat our foes to it: literally a matter of national
survival. Recombinant DNA research--taking its promises at their
most lavish--is not in the breathless battle to avert disaster,
but in the long haul to improve man's lot in any number of single
respects. If the new opening for one among them which science is
now probing turns out to be a boon, it will be one which until
recently nobody could yet expect and surely mankind had not been
straining for with a desperate need. And if, for once, the impa-
tience of the scientific quest for its own satisfaction finds
itself reined in by the presumptive beneficiaries of its fruits,
it is not the strawman of 'antiscientism' which should be knocked
down for it. The unknown as such, not irrational fear, enjoins
circumspection all the more as we can well afford its minor delays.
Move into the opening so conclusively demonstrated we shall, no
doubt, for science simply does not close a door again; but we can
move into this without the furies of necessity at our backs. If,
after a holding interval of (perhaps) overcaution, the potential
dangers turn out to have been exaggerated, neither science nor the
common wheal will have suffered a grave loss.

IV. BEYOND RESEARCH AND INTO THE OPEN WORLD

A certain narrowness of the preceding discussion should be
noted. In keeping with the terms of this paper, it was confined
to the *research* phase of the incipient story of recombinant DNA—
the only one actual at the moment. There, inadvertent "escape"
of potentially harmful laboratory creations is the danger to be
guarded against, and this is still a matter of human diligence at
the source, for which techniques exist, and which can be buttres-
sed by law. We have assumed that the odds against misadventure
can be made credibly high. But what will happen, once research
has done a successful job and passes its product on to outside
use, is another matter entirely. With massive industrial use of
bacterial populations, "containment" becomes virtually impossible;
and some of the anticipated uses even entail outright seeding into
the open environment (soil or ocean). The possible impact of such
viable microbial newcomers on the ecosystem is wholly unpredictable; it is past control, and possibly irreversible. We would
gamble with the unknown. Perhaps, therefore, these novel agents
should never be released at all, whatever their proximate advantage. With this broader question, which leads outside the
research context, we are here not concerned, though it looms
behind it as a much graver issue than freedom-versus-safety of
inquiry.   But the likelihood of its arising casts a paradoxical
light back on the research enterprise itself. If we conclude (as
I think we should) that planetary responsibility forbids that
gamble, which no dire necessity dictates, and join with that conclusion what we know of the inevitabilities of a free enterprise
system—then Science should be enjoined precisely *not* to aim at
potentially "useful" organisms in its recombinant experiments,
but to seek the *theoretical* enlightenment it professes to seek
from combinations guaranteed to tempt no one: one is as good a
teacher as the other about the basic mechanics of life. This, I
daresay, sounds like perverse advice in our age and day—"avoid

the useful!"--but as far as the free search after *truth* is concerned, removal of the practical lure could well redound to the
purity of that search. Certainly, the scientist qua scientist, if
in earnest about his care for truth alone, has here little to pro-
test. However, I well know the idea to be quixotic in the given
atmosphere of scientific endeavor, though even more quixotic is
the thought that at any later point things could still be controlled.

At the end, I return to the research side of the recombinant
syndrome with one more observation. The deliberations about
routine matters of safety and caution, which the microbiotic objects
of the actual research suggest, must not make us forget that
recombinant DNA research may well ring in an even more momentous
new chapter in man's dealing with living nature, including his
own. In sticking to the microbiotic program of the new research,
which is underway and will go ahead, we have stayed away from its
possible macrobiotic application, which opens questions of much
more than caution. About them I will here only say that, when it
comes to genetic manipulation on this plane, in particular to
changing human types or their hybridization with alien genetic
material, questions not of incidental risks but of essential admissibility arise whose responsible answering will require the ultimate wisdom of the race--far beyond mere prudence and certainly
beyond the special competence of scientists and their rights to
free inquiry. No threat to the ecosystem or to human health would
be involved, but something impalpable of transcendent rank and
therefore much harder to defend against the adventurousness of
the emancipated mind. One trembles to think of the residues of
reverence on which the race will be able to draw when it will need
a countervailing force against the heady lure of "creative" experiment with its own image.

Beyond Recombinant DNA—Two Views of the Future

Burke K. Zimmerman

Subcommittee on Health and the Environment
Committee on Interstate and Foreign Commerce
House of Representatives
Washington, D.C.

It is perhaps incumbent upon the author of the final commentary in a volume such as this to extract from the totality of the five-year history of the recombinant DNA debate, a sense of direction. As if told in advance I would have the last word in an argument, it now becomes not a question of trying to impress my point of view and set of values upon the minds of the participants, but requires a degree of detachment--a perspective from which the important lessons may, when not obscured by rhetoric and detail, become visible. Thus, I will try in these pages, not to resolve the multitude of uncertainties surrounding the use of a technique which can alter the genetic make-up of living organisms, but to analyze the nature of the debate itself. It is a debate among human beings. And it is human beings who will ultimately determine the effect that the recombinant DNA phenomenon will have upon our future. Thus it is from this perspective that I shall attempt

to extrapolate the events of the present to the ways in which
future societies may be expected to conduct themselves.

I will present two of the many ways one might choose to predict what lies ahead. One is based upon the observation that our
methods of dealing with any complex or controversial issues are
intimately tied to innate human behavior. The other concerns our
capacity for social and cultural evolution in dealing with the
problematic questions of the power of knowledge and the control of
technology. Neither view is particularly optimistic. But, I am,
by nature, an optimist, and continue to cling to the hope that
analyses such as these might contribute some small amount to
society's ability to contend with a set of increasingly difficult
problems.

I. ON HUMAN BEHAVIOR

Some five years ago, a remarkable event occurred, unprecedented
in the history of science. Almost as soon as the recombinant DNA
techniques for inserting new genetic material into viruses and
living organisms were established as proven and relatively simple
laboratory procedures, scientists took the initiative in calling
attention to the uncertainties which accompanied the use of these
methods. A voluntary moratorium on certain classes of experiments
was called, and a long series of deliberations was undertaken by
scientists which resulted in the development of the National
Institutes of Health *Guidelines for Research Involving Recombinant
DNA Molecules*. It appeared that a most unusual opportunity for
science had presented itself. This time, because of foresight and
prudence on the part of those who would develop and use this
technique, any potential for disaster or public risk would be
fully evaluated and the appropriate safeguards taken where needed.

Science, or at least technology (the important distinction
will be discussed in part II), had been under criticism for

rushing ahead with the applications of new scientific knowledge
without giving much thought to the disastrous consequences which
might ensue (*e.g.*, nuclear energy, chemical pesticides). Perhaps,
given the hard learned lessons of the past, science would emerge
in the 1970s with a new sense of social awareness, responsibility
and public accountability. It seemed that, when viewed fifty or
a hundred years hence, the human race would indeed be proud of the
wisdom of its forebears, its handling of the new technique of
gene-splicing standing as a milestone in cultural evolution.

At this time, however, some five years or so into the recombinant DNA era, I would say that there is little hope that we will
look back upon these years as a monument to the rationality of
humankind. That interest in this topic would cease to remain the
exclusive domain of science was a surprise to no one. The
discussion, which started out so promising, with the assurance of
public and environmental safety as its main concern, has taken so
many unpredictable new turns, that a wide variety of other issues,
all of them controversial, have been dragged into the debate (and
indeed it has become a debate), many of which not only transcend
the questions of safety in using the gene-splicing technique but
often have nothing to do with it at all. The scientists now com-
prise a minority of those expending their energy on this topic,
and those who do, spend much of their efforts debating nonscientific issues in which they have little experience or expertise, such as the appropriate degree of public participation in
determining policy, the proper role of the government, the distinction between guidelines and regulations, whether or not legislation is appropriate, whether or not Federal standards should
override state and local laws, public disclosure and patents. In
fact, while still discussed occasionally, the scientific aspects
and the safe conduct of research have become topics which no
longer hold nearly as much interest as the variety of once-
peripheral issues that have now become central.

But, it is not simply the great expansion of the scope of the
discussion that has changed over the past five years. More significant, and perhaps a corollary to the fact that numerous segments of society are now involved, is the substantial change in
character of the discussion. In the early years, at meetings like
Asilomar, the nature of the dialogue was much like any other
scientific meeting--not always orderly and logical, occasionally
emotional--but relatively focused upon a single issue.

Now, even those who represented the calm voices a few years
ago number among the many vocal participants in a highly polarized, emotional and generally inarticulate controversy. While not
without adversarial elements in the beginning, it has now
degenerated from a gentlemanly discussion into something of a
brawl. The voices of reason, some of them anyway, still persist.
but few are heard. And volumes like this one, full of profound
scholarly dissertations, will be all but ignored, at least for the
present.

Why has this happened? Clearly, controversial public issues
are nothing new. But this seems to be different than most. Has
it simply become too complex and difficult an issue, with too many
uncertainties to be dealt with competently and rationally by *any*
group of people? For all the talk of "public participation," this
has been a debate of the elite segments of society, the property
of the intellectuals. A quick survey of the Table of Contents in
this volume will attest to that. Moreover, almost every aspect of
the issue is being argued in the abstract, except, perhaps, by the
scientists actually performing the experiments. The hazards are
speculative, and the benefits to society are hypothetical. While
stories of laboratory-created monsters have held some degree of
appeal for those seeking a vicarious escape from their humdrum
lives, it is unlikely that recombinant DNA will ever capture the
serious interest of a significant part of the population. Even as
a proper subject for cartoons in the *New Yorker*, it is a concern
of the elite. It simply lacks the usual ingredients for a good

populist issue--or even a marginal one. There are no real economic considerations--jobs, wages and taxes are not involved. Nor
are consumer goods and services or any other parameter affecting
the quality of life. It is not really even an environmental
issue, in that the risk of some environmental disruption is considered remote even by the severest critics of the research. As
a public health issue, particularly in the absence of any documented case of the creation, through gene-splicing, of a novel
virulent pathogen, let alone infectious cancer, the general public
is simply not interested. Given the level of public apathy which
exists concerning other insidious hazards to health--exposure to
radiation from medical x-rays, environmental pollutants and even
smoking--it is small wonder that most people neither know nor care
about recombinant DNA.

It is relatively easy to understand why matters which might
directly affect a person's life and well-being are generally
dealt with by the public simply and usually very emotionally.
Overall, the abstract implications of an issue are given little
quarter and only the immediate future matters. Few would dispute
the notion that human beings are, after all, simply another
species of animal, slightly advanced in some of its intellectual
functions, perhaps, but a species, nevertheless, whose primary
motives and drives are instinctive and fundamentally irrational.
The greater capacity of human beings to reason allows for greater
success in satisfying these basic drives and needs. Detached
contemplation and abstract reasoning about matters which do not
affect one's immediate needs and desires are a product of civilization, and of the creation of leisure time, at least for a few.
But philosophy, science and other scholarly pursuits have always
been limited to a small intellectual class in every society, and
ours is no different.

How, then has it come to pass that an abstract discussion
engaged in almost exclusively by the elite elements of our
present society has become so fraught with bitterness and hysteria

and is anything but an example of the application of reason and
logic to what has become a complex issue?   At least, that is what
it appears to be from my Capitol Hill vantage point.   The answer,
I think, lies in the fact that intellectuals, scholars and
scientists still are human beings with the same basic animal
drives, instincts and needs as anyone else.   While the issues may
not revolve around the most primitive of human needs, they do,
nevertheless, involve areas of great concern to the various participants in what has become something of a verbal free-for-all.
I contend that what has happened is that a large fraction of those
in the middle of the fight are there because they perceive themselves as being threatened.   Thus the response we are seeing is
the primitive animal response to a threat: fear.

The case is not a difficult one to build.   When we examine the
various factions in this grand debate, the nature of their arguments and what they are objecting to or asking for, the conclusion
is inescapable that much of the heat of the discussion is generated by a fear reaction accompanied by defensive action and,
occasionally, counterattack.   On that level, it is a confrontation
and fight in the most classic sense.   At the risk of some of my
good friends taking it personally, it is at times reminiscent of
films I have seen on the social behavior of tribes of baboons.
While the topics of controversy may be somewhat loftier, the
threats, fears, actions and reactions are rather similar.

At this point in the discussion, it is only fair to point out
that there are many individuals who have either risen above their
baser motives or are personally detached enough from the central
issues, or are otherwise sufficiently enlightened and secure
people to have remained objective and rational throughout.   And,
perhaps for these very reasons, such individuals do not see themselves as personally threatened and approach the fracas with a
measure of calm rationality.   But, therefore, they also do not
have the degree of personal commitment to the struggle as those
who are primarily reacting rather than reasoning.   Hence, it is

the latter group--the emotional primitives if you will--who talk the loudest and longest, and who are intent upon making all bend to their views and upon shaping public policy accordingly. It is these whom one hears and sees the most, and, thus, those who have caused the debate to be characterized by their own human weaknesses.

I must admit that my perspective, as one charged with the responsibility of developing legislation on recombinant DNA, may be one which has forced me to see primarily the most outspoken, dedicated and personally involved participants who wish to influence Federal policy, and, therefore, those whose reason is applied toward achieving a previously determined end. The sea may well not seem to be nearly as stormy from other ships. And, even from Capitol Hill, the tempest seems to be subsiding somewhat.

The current debate might indeed be compared to a dying super-nova, fading in intensity as attention spans run out and people lose interest. The origin of the explosion can be traced to the time when the discussion on recombinant DNA spread beyond NIH and escaped the confines of the scientific community. It really began to take on significant proportions at the time of the open meeting of the expanded NIH Director's Advisory Committee, in February 1976. At that time, representatives of a number of public interest groups and professions outside of scientific research were invited to participate in one way or another. Recombinant DNA was a technically difficult subject for non-scientists to grasp immediately, although the practical implications of such a method were apparent. Thus, the degree of caution and concern on the part of many learning about it for the first time was understandable. But, to confuse the minds of the non-scientists further, it was clear that there was substantial disagreement among scientists themselves. Thus, even to seek a guru for advice became a difficult task. Some of the scientists, of course, were more credible than others, but non-scientists could not readily distinguish the most logical and rational arguments

from those laced with unjustified assumptions, a situation which
persists today. The tendency, therefore, was for people to form
opinions based upon the particular scientist's views which rein-
forced their own biases, fears or subconscious desires.

Then, the magazine articles began to appear, and the journal-
ists tended to play up the more novel, sensationalistic aspects of
the recombinant DNA phenomenon. And, in the wake of movies like
"Jaws," "The Andromeda Strain," "Earthquake" and many others,
there was nothing quite so provocative as a good disaster
scenario. That is not to say that some of the concerns expressed
were not legitimate ones. However, by playing up the science
fiction melodrama and not the fact that NIH was developing rather
conservative safety standards for the conduct of gene-splicing
research, there was fear engendered in many outside of the scientific community that indeed there was a danger, even an imminent
danger, of carelessly created, genetically modified living organisms which would perpetrate all manner of horrors. One of the key
articles, appearing in the *New York Times Magazine* in August,
1976, was, in fact, written by a scientist and immediately aroused
the fears of a much broader segment of the population.[1] It was
still an elite public. Who, after all, reads the *New York Times*?
This article, labeled by many as fear mongering, was read by
many members of Congress and the Senate, for whom this was their
introduction to recombinant DNA.

Throughout this period, the fact that the NIH Guidelines
applied only to NIH supported research was troublesome to almost
everyone. While there was much disagreement over the degree of
stringency of the Guidelines, and the procedures by which
decisions were reached, few disputed that *all* research, however
it was supported, should be subject to the safety standards in
these Guidelines. And, given the less than pristine record of

---

[1]See Cavalieri (1976).

some of the drug companies in the testing of new drugs and their
compliance with FDA rules, extension of the Guidelines was, for
many, the most important issue. For all the other controversies,
this notion was never in dispute, although the means by which
this should come about has been the focus of much of the disagreement.

Thus, while an interagency committee of the Federal government
was in the process of deliberations which would recommend that
legislation was the only proper way to extend the scope of the
Guidelines, some members of the U.S. Congress were drafting bills
to protect the American people against the potential hazards of
the creation of genetic chimeras. By this time, the debate was
well out of hand.

In order to illustrate my general observations of the preceding pages, I would now like to consider in some detail the
conduct of four of the most visible and vocal groups among the
many protagonists. I ask the readers' forgiveness for occasional
generalizations, for which there are certainly many exceptions.
The groups are the U.S. Congress, the so-called "public interest"
groups, the scientists, and the universities. I shall begin with
the Congress, although it is a diverse and difficult entity from
which to draw a stereotype.

The Congress

The motives of at least some members of the legislative
branch of the Federal government (who admit that the *New York
Times* article had a great deal of influence on them) would have
to be construed to include a substantial degree of fear, reflected in legislation which went beyond caution and a legal extension of the authority of NIH Guidelines or the equivalent. There
was more than just the fear of hazard for lab workers, an epidemic of infectious cancer and cataclysmic ecological disruptions,

but a fear--or at least a mistrust--of science and scientists, and of Federal agencies acting in conspiracy with science to put one over on the public, or at least avoid being responsible and accountable to the public for its actions. These fears may not have been totally without foundation, but because science was *perceived* as a public threat, the response was fear and thus an overreaction to what may have been factually justified. So bills were written, or at least amendments to bills proposed and some adopted, which were very defensive in character. Not all Congressional bills were of this nature--some were structured along the models of most current regulatory legislation, which authorizes a government agency to promulgate, administer and enforce regulations. Nearly all bills intended that the safety standards would be those in the NIH Guidelines--this was not a significant point of contention within Congress. But the degree of procedural detail, administrative specificity and bureau-cratic complexity differed greatly from bill to bill, varying according to the level of fear and mistrust in its sponsors.

Of course, it is impossible to categorize Congressional opinion. The degree of legislative complexity and the nature of various provisions often reflected the degree of sympathy a particular member of the House or the Senate had with the more stereotypable groups' positions discussed below.

The Public Interest Groups

The public interest and environmental groups--at least those which have taken an interest in the subject--have generally favored rather strict controls but have not really been able to find too many grounds upon which to judge the NIH Guidelines as too lax. But the most outspoken members of these groups-- Friends of the Earth, the Natural Resources Defense Council, the Environmental Defense Fund and the Sierra Club--tend to harbor

the same fears that resulted in the most severe legislative proposals. That is, I detect the suspicion that science and NIH are
in collusion to deceive and mislead the "public," or at least
exclude the public from the decision-making process so that it can
pursue research with neither constraints nor public accountability.
In the collective minds of many of these critics, the scientists
are stereotyped individuals, dripping with hubris--the quality of
nasty, self-serving arrogance discussed so eloquently by Lewis
Thomas in a commentary in the *New England Journal of Medicine*
(see Thomas, 1977).

Because of these fears, such groups, along with sympathetic
members of Congress, have called for very detailed legislation,
because they do not trust DHEW to do things right--*i.e.*, in the
public interest--if not told explicitly by Congress what to do.
The emphasis is most often on "public participation." That means
that the world contains basically two types of people--scientists
and the public who coexist as adversaries (and this dichotomy is
always drawn). Just who the public consists of is not usually
specifically defined, but there must be "public" members on
statutory advisory committees, regulatory commissions, biohazards
committees and study commissions. The assumption is made that if
a variety of non-science professions are specified, then decisions
will better reflect the public interest. The proposition that
diversity in the memberships of policymaking bodies is desirable
is entirely defensible on a number of grounds. But it is often
based upon the tacit assumption--the fear, if you will--that all
scientists think alike and will always vote in their own self-
interest rather than that of the public. That is, scientists are
not part of the public.

As a scientist, I have a great deal of trouble with that
notion. While I am generally very sympathetic to most environmental and public concerns, I simply cannot accept such a
generalization. There are, to be sure, a few scientists who are
thoroughly imbued with self-righteous arrogance, and give their

profession a bad name. Some of these have been rather outspoken
on recombinant DNA. And the small number who have openly flouted
the Guidelines have made matters potentially very bad for their
colleagues. But is it really so different from the self-
righteous arrogance of some of those who claim to represent the
public interest, and, therefore, who proclaim their views as
dogma? The arrogance of wearing the halo of self-proclaimed
purity and goodliness occasionally becomes too much to bear.

A list of basic--or perhaps "base" would be a better word--
human qualities, headed by fear but including mistrust, arrogance,
and self-righteousness--none of which is conducive to a rational
approach to *any* problem or controversy--would appear to govern
the performance and behavior of many of the so-called public
interest groups. There is often a group mentality expressed which
is greater or more extreme than that of the members of the group.
Together, they may reinforce each other's biases and fears and
take a collective position which an individual member alone may
have found too extreme. But, then, this "bandwagon" effect has
shown up in all groups of protagonists.

But, many other human failings are tolerable as long as individuals are willing to learn and keep their minds open. Failure
to do that has to be the greatest sin any debater can commit. Yet
few of the public interest people have really tried to understand
the recombinant DNA technique and review the scientific basis for
the presumed hazards. Nor have they really tried to understand
the profession of science or what motivates a true scientist.
If there could be true communication and an attempt at understanding, rather than an intransigent designation of the "enemy,"
the level of the debate would be raised considerably.

Before leaving the public interest groups' failings, it is of
interest to point out an irony which still baffles me. One of
the complaints of many environmentalists about the rule-making
procedures within the government is that they are terribly time-
consuming, complicated, expensive, inflexible and still often

ineffectual. The process through which EPA goes to promulgate
an effluent standard for a single toxic chemical may take two
years. How the bureaucracy became so complex and slow is unclear,
but it often strikes us as a preposterous waste of taxpayers'
money when the job could be done just as well or better with far
less fuss. Yet the types of legislative proposals endorsed by
the public interest groups are even more innately bureaucratic
than most existing regulatory legislation and could not only create
administrative nightmares for the responsible agency, but nightmares for those who would wish to see the rules changed. They
have even picked up the jargon of Washington. A lobbyist for one
of the most active groups once suggested that I should write into
the DNA bill a provision for an interagency committee to
"prioritize" a list of risk assessment experiments.

The Scientists

While the fears of some Congressmen and environmentalists
have provided us with a demonstration of mankind's fundamental
irrationality, that exhibited by a great many scientists was perhaps the most difficult to contend with. There have been a few
fights over levels of containment within the Recombinant DNA
Molecule Program Advisory Committee which developed the Guidelines. However, the Scientific community generally accepted the
NIH Guidelines and endorsed the notion that the standards therein
should be applicable to all recombinant DNA activities. A few
molecular biologists denounced the Guidelines as silly because,
they said, no harm could ever possibly come from a recombinant DNA
experiment. Such individuals, however, were and are in a small
minority.

Nevertheless, the greatest fear response exhibited by any
group came from the scientists as soon as legislation was proposed.
It was particularly frustrating for me to deal with a barrage of

protests so fraught with a nearly total lack of understanding of
administrative law, often a lack of knowledge of the content of
particular bills and a failure to distinguish between the various
House and Senate bills. The extent to which bills were misunderstood, misinterpreted and false conclusions drawn from them was
unbelievable. One might not expect scientists to be adept at
reading and understanding Congressional bills--after all, most had
never seen a bill before they got hold of the first DNA bills. To
be sure, there were some features in some of the bills which were
legitimate concerns. But the fear response was nearly as intense
for the bills which did little more than apply the Guidelines uni-
formly.

The most offensive feature of this reaction of scientists was
not their initial ignorance and naivity--that can be forgiven--
but their subsequent refusal to learn. Numerous briefings were
held and memoranda written to explain in detail how each section
of the House bills should be interpreted, but a significant segment of the scientific establishment resolutely held steadfast to
their misconceptions and false conclusions. This was something
worse than hubris and basically unforgivable.

Therefore, since these are, after all, intelligent individuals
well trained in a system of logic known as the scientific method,
one must conclude that this was purely an instinctive, emotional
and defensive response to fear, and as such, any attempts to dis-
suade such people by appeal to reason would be predictably useless. But fear of what? How could the mere extension of safety
standards by law pose such a threat?

Clearly, if the purpose and content of legislation had been
understood in the first place, it wouldn't have been perceived
as a threat at all. But since it was somehow regarded as *control*
of the content of scientific research, where scientists were to
be sent to jail for forgetting to plug a pipette, no wonder such
a frozen state of emotional intransigence resulted. The totally
spurious issue of the freedom of inquiry was raised, again

through errors of perception of actual legislative proposals.
Nor was the fear of setting a dangerous precedent for the regulation of science a real issue, since none of the bills either prescribed or proscribed the content of scientific research.

I almost sensed that the rallying cry to stop legislation was
not unlike the student strikes of the late 1960s and early 1970s.
I once asked a coed at a demonstration, whom I recognized as being
a student in a class I was teaching at the University of California at Santa Cruz, why she and her friends had just taken over
the Administration building. Her reply, "In sympathy for the
People's Park issue in Berkeley, to show this University our
grievances and for revolutionary principles in general," might,
in spirit, be much along the lines of the answer I would get from
many scientists opposing legislation. There is a strange
exhilaration in getting involved in a cause. Few would question
that it is a lot more exciting than the daily routine of giving
your lecture to your class of sophomores, going into your lab and
reviewing the day's work with your technician, reading your mail,
correcting exams, and reviewing graduate student applications,
day after day after day.

A petition opposing legislation was circulated at the Gordon
Conference on biological regulatory mechanisms last year. Signed
by most of the scientists at the conference, the petition was
sent around to many universities and research institutes for sig-
natures, collecting some 900. From reports on the Gordon Conference, it was clear that the petition had been written and
promoted by a very small number of individuals suffering from the
self-righteous indignation of imagined threats. In fact, the
covering letter to Congressman Rogers from Fred Blattner accompanying the petition contained a statement embodying a still
widely held fallacy, referring to ". . . the opposition to the
regulatory approach embodied in H.R. 7897 in view of the speculative nature of the hazards involved." The fallacy is, of
course, the assumption that the legislation is a response to the

hazards. Rather, it is the NIH Guidelines, already administered
to NIH grantees, more stringently than most actual regulations,
which are the response to the speculative hazards. The legislation, specifically the above numbered bill, was a response to the
fact that the Guidelines did not apply to research beyond the
auspices of NIH. Legislation then, as now, is justified only to
the degree that the NIH Guidelines represent sound policy for dealing with speculative hazards. Why, then, didn't Blattner's letter
or the petition address the stringency of the Guidelines? The
Guidelines are the substantive controls, not the legislation.

Perhaps the answer lies in the misunderstanding of the legislation revealed in the petition itself. There is a tenuous statement of endorsement for the NIH Guidelines. But the first mention
of the bill is to quote a section of the "findings" out of context, which was included in the bill not as an editorial response
of Congress to recombinant DNA research, but to justify the con-
stitutionality of the proposed legislation, in this case under
the Commerce Clause. The findings do not appear in the final
statute.

The petition, making no distinction between Kennedy's S. 1217
and Rogers' H.R. 7896, goes on to allude to "inflexible and
unwieldy bureaucratic machinery" which will "severely inhibit
the development of many fields of knowledge," "prior restraints
on scientific inquiry" and states how society will be deprived of
the benefits of such research if legislation is passed. In view
of the fact that, at least under H.R. 7897, the NIH grantee would
scarcely notice any difference in the way the standards were
administered, the series of non-sequiturs in the petition is most
curious. I can only conclude that if those drafting the petition
started out by believing such nonsense, then the state of fear of
legislation must be such that they are in no condition to listen
to reason, or to a simple explanation of the facts.

The danger in the errors of a few is that such individuals
may be persuasive enough to spread their errors of perception and

reason along with their fears to large numbers of others. Noting
many familiar names among the nine hundred signators, I called a
few to see why they had signed the petition. Few had read the
actual legislation, and none could remember which bill had the
regulatory commission nor did it seem to matter much. One
acquaintance remembered that he had signed the DNA petition but
didn't know whether it was for legislation or against it. He had
to go check with a friend who had been at the Gordon Conference
to refresh his memory.

The big guns of science were some of the most outspoken
critics. Arthur Kornberg asserted in a letter dated July 22,
1977, to Frank Press, the President's Science Advisor, that "even
the most benign legislation under consideration will seriously
impede basic biologic research and the spirit of free inquiry."
Terms like "Lysenkoism" were dragged into the argument and the
lines were drawn for a real battle. Not all scientists supported
the strong words of the petitions or agreed with Dr. Kornberg,
but most were antagonistic to legislation to some degree.

Probably the worst sin committed by the scientists, however,
was the misuse of and, in fact, overstatement of scientific data
to further a political view. After the Falmouth risk assessment
workshop, in which data were presented which showed it highly
unlikely that $E.$ $coli$ K-12 could be converted into an epidemic
pathogen, statements were rampant to the effect that "the risks
have been shown to be much less than previously thought," imply-
ing that the new safety data applied to all host-vector systems.
In fact, data are available $only$ respecting the pathogenicity of
$E.$ $coli$ K-12. A similar misuse of science occurred with the
release of a manuscript by Stanley Cohen showing certain types
of site-specific recombination could be made to occur within an
$E.$ $coli$ cell under very specific laboratory conditions. Yet the
experiments were cited as "proving" that nearly all types of
recombination, even between bacterial and eukaryotic DNA, occur
in nature.

It is clear that such conclusions would never have been tolerated in a responsible scientific forum. But in the emotionally charged atmosphere that resulted from the fear of legislation, highly competent and usually responsible scientists were allowed by their equally able colleagues to say outrageous things, at apparently minimal risk to their credibility. The spirit of the group crusade against what was perceived as the forces of evil, as well as a very strong "bandwagon" effect combined to carry at least a significant and highly visible portion of the scientific community far outside the boundaries of calm rationality.

The Universities

One of the most intensely debated issues of the recombinant DNA controversy is that of whether or not Federal requirements should preempt state and local regulations. Most will argue that a uniform Federal Standard makes sense. But the public interest groups, and some others, argue that the right to pass stricter state and local rules is an expression of "public participation," perhaps the *only* chance the public has to get directly involved in the issue. The Cambridge, Massachusetts debate and hearings were the beginning of the focus on this one issue. It rallied the public interest groups and a few scattered local citizens' groups which, for the most part, were again reacting to fear. There was either a university or a drug company "doing that stuff and God knows what might come out of there." This is as close as it has come to a public issue—but it still may be a function of a statistical fringe which reads the *New York Times* and needs to have a local cause. It has still only arisen as an issue in a handful of communities and a few state legislatures.

Naturally, if the local citizen leaders are responding out of fear, they pose a threat to the universities and research

facilities in the political jurisdictions at issue. And these
institutions, predictably, will respond, again out of fear, to
make sure Federal laws override all state and local laws.

A good case in point is Harvard, the leader among a small
group of universities which have engaged in an intense lobbying
effort to have whatever Federal legislation is passed have strong
Federal preemption language. Their position has been close to
indifference on all other aspects of the bills. But the Cambridge city ordinance must be overturned at all costs. Minor
word changes in preemption language at mark-ups are perceived as
betrayal and sell-outs by the preemption lobbyists, so intensely
loaded an issue has it become.

Frankly, I must regard it as a third rate issue. I would
expect that as soon as Federal standards are passed, there will
be little incentive for state and local action. But the blood
still runs high among the university and public interest lobbying
groups. The preemption fight has now lost most semblance of
reason or of proper perspective in its relation to the rest of
the recombinant DNA issues.

Given the immense hassles of 1977 and the toll taken by all
of the groups discussed above in chewing legislation to shreds,
it appeared that another means had to be found to achieve the
original objective of extending the safety standards of the NIH
Guidelines to all parties. Therefore, I have generated yet
another bill, H.R. 11192, which at this writing, has just been
reported by the House Committee on Interstate and Foreign
Commerce. This one is a two year interim control bill, and is
about the simplest means to accomplish this goal in a way which
is also enforceable, administratively sound and flexible. Most
of the administrative details written into earlier bills have
been eliminated.

The proposed legislation was designed to have a chance of
squeezing its way through the barrages of attacks it has already
received and will certainly continue to get from the public

interest groups and their allies in Congress. As predicted, it
is providing a battleground for what promises to be the most
vicious fight yet over preemption. I am reminded of the movies
where the king of France is squaring off against the king of
England, with rows of multicolored horses and soldiers lined up
on opposing hillsides, waiting for the signal. With the introduction of the bill, the charge has begun.

Should it survive the gauntlets of Congress and eventually
become law, it will be in spite of the perversities and illogic
of the protagonists. If it does not, then I shall have to concede victory to the forces of darkness. A massive collection of
fear reactions, a failure and unwillingness to understand all of
the necessary elements in the debate, and a breakdown in effective
communication will have combined to have done it in, as they did
last year. Ironically, if such is the case, it will only have
stopped what nearly everyone wanted all along anyway.

Is one to conclude from this that the issues were simply too
complex for any set of human beings to deal with all of the
parameters at once in a way that makes sense? Or is it the fact
that the irrational components of human behavior are simply so
great as to prevent even highly trained, intelligent individuals
from dealing effectively with such a difficult set of subjects?

I suspect that both are true. I am not particularly
encouraged by the level of imagination or wisdom exercised in
dealing with complex problems by either ancient or contemporary
civilizations. Perhaps, ultimately, cultural evolution is
possible to a degree which may lessen the difficulty of society's
tasks. But, if Alvin Toffler is right, the technological complexity of our society has already surpassed the level of our biological and social evolution to deal with it effectively.

More true renaissance men and women would help. Part of the
difficulty I think, is that we have become much too much a
society of specialists. The degree of technological sophistica-
tion of our society may demand specialization. But if we are to

ultimately survive it, it may also demand generalization. People who can transcend the specifics to deal with the more global issues which span many specialties are needed. It is perhaps little wonder that an issue which included molecular biology, infectious diseases, administrative law, ethics, philosophy, and probably should have included psychiatry, could not be dealt with more effectively than it was.

I have no formulae to prescribe. The faults seem to lie in innate human nature itself. It may be possible for cultural evolution to succeed where biological evolution has not, but I wouldn't count on it. Perhaps the only cure would lie in the ultimate irony: to genetically engineer the human race to keep it from destroying itself by expressing its "humanness."

II. ON SCIENCE AND TECHNOLOGY

The nature of the recombinant DNA debate, at least at the level where public policy is made, was discussed in the previous section. It might be considered to be a micro-example of the entire democratic process, in which brilliant, creative public policy decisions are virtually precluded. To get even minimal or compromise legislation through Congress is usually to be considered a victory of sorts. But, for the same reasons, the process also precludes most truly egregious public policy decisions.

Nevertheless, from time to time, perhaps due simply to statistical fluctuations, the government on rare occasions makes decisions worthy of high praise, such as establishing the National Institutes of Health, as it makes ones which are unforgivably bad, such as the policy of pursuing the war in Viet Nam. There has been a hint of what could perhaps have been or may yet be such a public policy disaster which has been present as a minor issue throughout the recombinant DNA controversy. It was raised partly out of fear or misunderstanding by some of those

individuals to whom the scientific aspects were new and baffling.
But, a small number of scientists and scholars also addressed
this issue, sometimes brilliantly. It is something quite apart
from the safety question--some of those who would put severe
restrictions on genetic transplantation research consider the
matter of containing potential hazards one which can be reasonably solved and for which the NIH Guidelines are indeed conservative. It has generally been outside of the heat of the battle
over legislation, and not generally subject to the perceptual
errors and refusal to understand which has characterized the
debate. It is not a new issue and, in fact, may be considered a
profound philosophical problem of a complex technological society.
Hans Jonas, in the preceding paper, also raises the following
question, and he can be very persuasive.

How should society and government deal with the emerging
power of science and the application of knowledge to new and,
perhaps, threatening technologies? There are many misperceptions
and misconceptions of what science and what technology entail and
what the distinction between science and technology actually is.
However, the real question that has emerged has been whether or
not society has the right to limit the acquisition of knowledge,
and, if so, under what conditions. This is not a new question.
It is perhaps only a revival of a very old question--a question
that, in fact, was answered in medieval days, during the
Renaissance and on several occasions in modern times by various
governments or religious leaders. There have been instances
where the truth is politically dangerous. There has also been
the fear that the truth might be used for purposes other than
those in the best interests of society or a particular government. The truth has often been proclaimed by law, by royal
decree, by papal edict, and by majority vote, regardless of what
the facts and observations may have dictated.

Nevertheless, the question has arisen yet another time.
Some of the participants in the recombinant DNA discussion have

suggested that society must control the study of certain subject areas and perhaps limit, or even prohibit, the acquisition of certain types of knowledge. The reason given is that the consequences of the technology which that knowledge might create at some future point might be too dangerous or too awesome for society to deal with. Or, to put it another way, they are suggesting that the acquisition of knowledge is not a sterile pursuit but that it is, in fact, intimately tied to ethical and moral considerations. It is argued that one should not simply pursue the truth without considering these ethical or moral considerations.

This is a contention with which I differ sharply. It is, of course, not always true that one can pursue scientific inquiry in a totally sterile environment. For example, violations of the human body, in order to gain knowledge, would have to be considered to be very much involved with ethical and moral considerations. In general, however, I would contend that the pursuit of knowledge *is* a sterile endeavor. Knowledge by itself is knowledge. Oh, it might require the use of certain experimental techniques in order to gain that knowledge and that is quite another question. That is, one must certainly demand that the techniques used in order to gain knowledge will not infringe upon other human rights, including the right not to be exposed to an involuntary risk to health. In fact, it is also reasonable to argue for the right of the environment not to be unnecessarily affected by the pursuit of knowledge. But these are situations where different constitutional rights, including certain human rights, may conflict. Clearly, we have to impose limitations on all of our rights in order to preserve all the other rights. That is not the question. The question is, is any kind of knowledge somehow tainted? Are there some kinds of knowledge that are good and some kinds of knowledge that are bad?

I would argue that it makes little sense to assign values to knowledge. To be sure, certain types of knowledge may be more useful in a practical sense, depending upon what social or personal values might prevail at a particular time and place. But knowledge per se is simply that—it is understanding. To say that some knowledge is good and that other knowledge is evil or that we should not hold certain knowledge because it may be used in an evil or unethical way makes absolutely no sense to me. It would appear that any knowledge, however innocuous it might seem, can, by some imaginative individual, be put to downright nefarious use. And, in fact, if we look at the history of mankind we see that the capacity for evil is immense—even where the factual or scientific basis of knowledge for perpetrating such has been minimal. Why, then, has that question been raised with regard to manipulating the genes of a living organism? Why is it more relevant than, for example, using an animal in experimental science in order to gain knowledge into living processes? Particularly, is it relevant when the organism being manipulated is a bacterium, with which relatively few people have some kind of emotional identity?

But this is, of course, not where the issue lies. The fear is that somehow there will be modifications of people, that perhaps some evil dictator at some point or perhaps even a scientist working in the secrecy of his own laboratory will begin to do experiments of an irreversible nature which have the most profound ethical consequences. The cloning of human beings has been raised as an issue which has been given a spurt of new vigor by the publication of a book claiming that the event has already taken place. That issue, however, has been a red herring for the most part, in that if and when science learns how to clone human beings it will probably not use recombinant DNA techniques. But what about genetic engineering in general? Suppose, for example, one could insert genes for high intelligence, so that the genetic distribution of a parameter which is considered to be desirable would be modified. Other "desirable" qualities could also be height, or strength, or good looks, or a woman's bust measurement.

They could be a great may things, if, indeed, it were possible to manipulate genes with this degree of control.

I am not going to even enter into the question now of whether or not the development of such a technology would be desirable or undesirable. For the most part, I do, indeed, see ethical diffi- culties in somehow modifying the human race. And the basis for these feelings are not rational. Like all things we "feel" strongly about, they are emotional and I won't attempt to defend them on any other grounds than that is the way I happen to feel. I would contend that society does have the right to control the use of knowledge--that is, technology--in directions which would be in accord with the collective social values operating at a given time. This right does not depend upon the basis for these social values.

In Congress, many people have contended that the legislation to provide uniform safety standards for the conduct of activities involving recombinant DNA should be a focus as well for these broader questions. It is asserted that we must not only provide safety for the conduct of recombinant DNA research, but consider as well the long-term implications of the uses of the technology, including the ethical considerations in modifying the genetic makeup of living organisms and address the whole question of public participation in determining the federal policy toward the conduct of science. It is, in fact, bringing up this latter issue that suggests that science is not sterile and that the "public," through mechanisms above and beyond our existing demo- cratic institutions, must say, or have a say in, whether or not it wishes to see a certain branch of science undertaken.

Because, historically and actually, the perceptions of scientists may be very, very different than the perceptions of the public, I am very disturbed by the notion that a public (a lay public, at that) should have the right to interfere with the questions a scholar is permitted to ask, or at least, the questions for which a scholar (and the term "scholar" includes

the term "scientist") may seek the answers. The notion that it
is more than an individual decision of an investigator is still
based upon fear. It is contended that science, or knowledge,
somehow has intrinsic values and that society is capable of
making the judgment of those values and therefore determine that
which it wants and reject that which it does not want. The
latter action--rejecting the acquisition of "bad" knowledge--can
only be considered the exercise of fear. This notion runs counter
to the concept of academic and intellectual freedom for which at
least a number of individuals in this country have been fighting
the past several decades.

Therefore, a fundamental and an important distinction must be
drawn between science--that is, the pursuit of knowledge--and
technology, which is the use and application of knowledge.
Clearly, the application of knowledge does have social consequences, economic consequences, and affects, ultimately, the lives of
all of us. And it is certainly in keeping with the notion of
government by the people that the public should have the ultimate
say in the directions to which a country's technological
resources are directed. Presumably, that would also mean that
society has the right to choose the development of a technology
which may be ultimately harmful to it in the long-run in order to
achieve something that's desirable in the short-term. In fact,
society has made many such choices. The choice to develop the
automobile as the primary mode of transportation, even though
it's depleting our natural petroleum resources at a tremendous
rate, is one such example. Sooner or later society will pay the
consequences for that. Nevertheless, this is a decision which
society endorsed and still endorses. Perhaps this also is an
illustration of the irrationality of mankind. Nevertheless, it
is consistent with our democratic principles. It is certainly
fitting, too, that society also has the choice to decide *not* to
develop those technologies which are perceived as harmful. And
it might well be good policy to prohibit at least for a time, the

genetic modification of human beings, except possibly under very
strict and well-controlled conditions. We already have a
national commission to discuss experimentation on human beings
and recommend rules. Certainly, the various questions of medical
ethics are constantly being raised and discussed. The ethics of
developing nuclear energy and even nuclear weapons may also be
properly questioned by society. But, society has made that
choice and must now live with a technology of great potential
harm.

Nowhere in history has the suppression of the acquisition of
knowledge ever been successful. It always happens simply
because there is fear of the consequences of the use of such
knowledge. If it is oppressed in one place, it will be pursued
in another, maybe not in the same decade or the same century but
certainly in the next one. Perhaps one of the most admirable
human traits that we see being pursued is the insatiable thirst
to understand the nature of the universe in which man lives.

In the evolution of society, there is a very important demand
placed upon human beings to learn how to deal with the awesome
power of knowledge. Perhaps the irrational components in human
nature and perhaps the shortcomings of our political and decision-
making systems are not quite on the same level with the state of
our knowledge. If so, it is entirely possible that ultimately,
through error or misuse of knowledge, we will simply destroy ourselves. Clearly, all of us hope this will not be the case. Does
that mean, however, that the ultimate solution is to suppress
the acquisition of even more knowledge because we cannot even
deal with what we have? No, I think not. Science, or study of
any sort for that matter, is fascinating, in part because of its
uncertain nature. It is inherently unpredictable. We do not
always know what we are going to find. And, in fact, in the
capacity for evil which new knowledge may possess, we may also
see the capacity for great good--here, good and evil being
defined purely in a social sense, of course.

So in pursuing knowledge we may indeed find the keys to get us out of some of the dilemmas that we are currently in, as well as the implements to create new ones. In any event, if we have learned anything from the recombinant DNA hassle other than to see that people have perhaps not evolved as far socially as we might like to think, let it be that we must not let our fear of the unknown stifle perhaps the noblest human qualify of all--that of curiosity.

* * * * * * * * *

I have presented for the readers two of my views of the future. One is a rather pessimistic outlook perhaps, because I have seen that by and large human beings have not dealt with the new recombinant DNA phenomenon with rationality, with wisdom or with foresight. Perhaps it is a result of this rather negative outlook that I am concerned that society could make a very serious and incorrect choice--a choice based upon fear. Hence, my second view of the future might be considered a caveat to all of us.

Uncertainty for many is very unsettling and troublesome. People would like to have a well-ordered society where everything is known and uncertainty does not exist. But frankly I find uncertainty and the fact that there are still vast areas which are not known or understood by human beings and the fact that our future is intrinsically unpredictable to be among the most exciting aspects of life. Life is not without risk, nor is it without uncertainty. The human race has been living with both of these qualities for many millenia. And it need not be uncomfortable to live with these qualities. But how may we better deal with them than we seem to be doing now?

Again, I must observe that a society of such technological complexity that we now live in has resulted in a society of specialists. People are so specialized that to find an individual capable of grasping the more universal problems or even caring to, is extraordinarily rare. Therefore, I close with a

plea that we train more true renaissance men and women, more
generalists, more philosophers, more well-educated people who, if
they must be specialists, let them be specialists in a variety of
diverse areas so that they may have some hope of seeing all of
the important parameters of a problem, not simply one, or two or
three. That, I would contend, has been indelibly underscored by
the debate on recombinant DNA.

REFERENCES

Cavalieri, Liebe F.   1976.   "New Strains of Life--Or Death."
    New York Times Magazine, August 22, 8-9+.
Thomas, Lewis. 1977.   "Notes of a Biology-Watcher: The Hazards of
    Science." *New England Journal of Medicine*, *296* (February),
    324-328.

Appendix

The decision of biologists to "go public" with their concerns over recombinant DNA research brought the debate out of the limited environs of a narrow academic specialty. Letters, published in *Science, The Proceedings of the National Academy of Sciences, Nature* and elsewhere, were the main vehicle for the communication of these concerns. Although initially these were directed to other members of the scientific community, the letters were quickly taken as the basis for many of the news stories covering these early developments. These letters themselves are important historical documents. The most common, and the dominant feature, of all of these letters is the genuine concern of the writers. This is an irreplaceable source for understanding the gradual shift in the attitude of the investigators themselves and in the overall tone of the debate. For these reasons several of them are reprinted in this appendix.

THE GORDON CONFERENCE

The decision to "go public" was made at the 1973 Gordon conference on Nucleic Acids. The co-chairpersons of the conference were Maxine F. Singer and Dieter Soll. Their letter appeared under the heading "Guidelines for DNA Hybrid Molecules."

303

September 21, 1973*

Those in attendance at the 1973 Gordon Conference on Nucleic
Acids voted to send the following letter to Philip Handler, president of the National Academy of Sciences, and to John R. Hogness,
president of the National Institute of Medicine. A majority also
desired to publicize the letter more widely.

We are writing to you, on behalf of a number of scientists, to
communicate a matter of deep concern. Several of the scientific
reports presented at this year's Gordon Research Conference on
Nucleic Acids (June 11-15, 1973, New Hampton, New Hampshire) indicated that we presently have the technical ability to join together, covalently, DNA molecules from diverse sources. Scientific
developments over the past two years make it both reasonable and
convenient to generate overlapping sequence homologies at the
termini of different DNA molecules. The sequence homologies can
then be used to combine the molecules by Watson-Crick hydrogen
bonding. Application of existing methods permits subsequent covalent linkage of such molecules. This technique could be used,
for example, to combine DNA from animal viruses with bacterial DNA,
or DNA's of different viral origin might be so joined. In this
way, new kinds of hybrid plasmids or viruses, with biological
activity of unpredictable nature, may eventually be created. These
experiments offer exciting and interesting potential both for
advancing knowledge of fundamental biological processes and for
alleviation of human health problems.
 Certain such hybrid molecules may prove hazardous to laboratory
workers and to the public. Although no hazard has yet been established, prudence suggests that the potential hazard be seriously
considered.

Reprinted with permission from *Science*, *181*, 1114. Copyright
1973 by the American Association for the Advancement of Science.

A majority of those attending the Conference voted to communicate their concern in this matter to you and to the President of the Institute of Medicine (to whom this letter is also being sent). The conferees suggested that the Academies establish a study committee to consider this problem and to recommend specific actions or guidelines, should that seem appropriate. Related problems such as the risks involved in current large-scale preparation of animal viruses might also be considered.

Maxine Singer

National Institutes of Health

Bethesda, Maryland

Dieter Soll

Department of Molecular Biophysics
    and Biochemistry

Yale University

New Haven, Connecticut

THE BERG COMMITTEE

The National Academy of Sciences responded to the Singer/Soll letter by asking Paul Berg to organize a committee to evaluate the research and to propose a course of action. The committee convened in April 1974 at the Massachusetts Institute for Technology. The scientific community was informed of the committee's recommendations in the following letter which appeared simultaneously in Science *and* The Proceedings of the National Academy of Sciences, USA *under the heading,* "Potential Biohazards of Recombinant DNA Molecules."

July 26, 1974*

Recent advances in techniques for the isolation and rejoining of segments of DNA now permit construction of biologically active recombinant DNA molecules in vitro. For example, DNA restriction endonucleases, which generate DNA fragments containing cohesive ends especially suitable for rejoining, have been used to create new types of biologically functional bacterial plasmids carrying antibiotic resistance markers (1) and to link *Xenopus laevis* ribosomal DNA to DNA from a bacterial plasmid. This latter recombinant plasmid has been shown to replicate stably in *Escherichia coli* where it synthesizes RNA that is complementary to *X. laevis* ribosomal DNA (2). Similarly, segments of *Drosophila* chromosomal DNA have been incorporated into both plasmid and bacteriophage DNA's to yield hybrid molecules that can infect and replicate in *E. coli* (3).

Several groups of scientists are now planning to use this technology to create recombinant DNA's from a variety of other viral, animal, and bacterial sources. Although such experiments are likely to facilitate the solution of important theoretical and practical biological problems, they would also result in the creation of novel types of infectious DNA elements whose biological properties cannot be completely predicted in advance.

There is serious concern that some of these artificial recombinant DNA molecules could prove biologically hazardous. One potential hazard in current experiments derives from the need to use a bacterium like *E. coli* to clone the recombinant DNA molecules and to amplify their number. Strains of *E. coli* commonly reside in the human intestinal tract, and they are capable of exchanging genetic information with other types of bacteria, some of which are pathogenic to man. Thus, new DNA elements introduced into *E.*

coli might possibly become widely disseminated among human, bacterial, plant, or animal populations with unpredictable effects.

Concern for these emerging capabilities was raised by scientists attending the 1973 Gordon Research Conference on Nucleic
Acids (4), who requested that the National Academy of Sciences
give consideration to these matters. The undersigned members of
a committee, acting on behalf of and with the endorsement of the
Assembly of Life Sciences of the National Research Council on this
matter, propose the following recommendations.

First, and most important, that until the potential hazards of
such recombinant DNA molecules have been better evaluated or
until adequate methods are developed for preventing their spread,
scientists throughout the world join with the members of this committee in voluntarily deferring the following types of experiments.

Type 1: Construction of new, autonomously replicating bacterial plasmids that might result in the introduction of genetic
determinants for antibiotic resistance or bacterial toxin formation into bacterial strains that do not at present carry such
determinants; or construction of new bacterial plasmids containing
combinations of resistance to clinically useful antibiotics unless
plasmids containing such combinations of antibiotic resistance
determinants already exist in nature.

Type 2: Linkage of all or segments of the DNA's from oncogenic
or other animal viruses to autonomously replicating DNA elements
such as bacterial plasmids or other viral DNA's. Such recombinant
DNA molecules might be more easily disseminated to bacterial populations in humans and other species, and thus possibly increase
the incidence of cancer or other diseases.

Second, plans to link fragments of animal DNA's to bacterial
plasmid DNA or bacteriophage DNA should be carefully weighed in
light of the fact that many types of animal cell DNA's contain
sequences common to RNA tumor viruses. Since joining of any
foreign DNA to a DNA replication system creates new recombinant

DNA molecules whose biological properties cannot be predicted with certainty, such experiments should not be undertaken lightly.

Third, the director of the National Institutes of Health is requested to give immediate consideration to establishing an advisory committee charged with *i)* overseeing an experimental program to evaluate the potential biological and ecological hazards of the above types of recombinant DNA molecules; *ii)* developing procedures which will minimize the spread of such molecules within human and other populations; and *iii)* devising guidelines to be followed by investigators working with potentially hazardous recombinant DNA molecules.

Fourth, an international meeting of involved scientists from all over the world should be convened early in the coming year to review scientific progress in this area and to further discuss appropriate ways to deal with the potential biohazards of recombinant DNA molecules.

The above recommendations are made with the realization *i)* that our concern is based on judgments of potential rather than demonstrated risk since there are few available experimental data on the hazards of such DNA molecules and *ii)* that adherence to our major recommendations will entail postponement or possibly abandonment of certain types of scientifically worthwhile experiments. Moreover, we are aware of many theoretical and practical difficulties involved in evaluating the human hazards of such recombinant DNA molecules. Nonetheless, our concern for the possible unfortunate consequences of indiscriminate application of these techniques motivates us to urge all scientists working in this area to join us in agreeing not to initiate experiments of *types 1* and *2* above until attempts have been made to evaluate the hazards and some resolution of the outstanding questions has been achieved.

Paul Berg, *Chairman*, David Baltimore, Herbert W. Boyer, Stanley N. Cohen, Ronald W. Davis, David S. Hogness, Daniel

Nathans, Richard Roblin, James D. Watson, Sherman Weissman and
Norton D. Zinder.

*Committee on Recombinant DNA Molecules Assembly of Life
Sciences, National Research Council, National Academy of Sciences,
Washington, D.C.*

REFERENCES AND NOTES

1. S. N. Cohen, A. C. Y. Chang, H. Boyer and R. B. Helling.
    Proc. Natl. Acad. Sci. U.S.A. 70: 3240, 1973; A. C. Y.
    Chang and S. N. Cohen, *Ibid.* 71: 1030, 1974.
2. J. F. Morrow, S. N. Cohen, A. C. Y. Chang, H. Boyer, H. M.
    Goodman and R. B. Helling, *Ibid.* In press.
3. D. S. Hogness, unpublished results; R. W. Davis, unpublished
    results; H. W. Boyer, unpublished results.
4. M. Singer, and D. Soll. *Science* 181: 1114, 1973.

THE ASILOMAR CONFERENCE

*The Berg Committee letter resulted in a voluntary moratorium
which was unprecedented in the history of science. Their fourth
suggestion, calling for an international meeting of involved
scientists, resulted in the Asilomar conference on Recombinant
DNA Molecules. The conference produced specific recommendations
for classifying types of experiments and for matching these to
levels of physical and biological containment. The following is
the "Summary statement of the report submitted to the Assembly of
Life Sciences of the National Academy of Sciences and approved by
its Executive Committee on 20 May 1975."*

ASILOMAR CONFERENCE ON
RECOMBINANT DNA MOLECULES*

Paul Berg, David Baltimore, Sydney Brenner, Richard O. Roblin
III, Maxine F. Singer

I.   Introduction and General Conclusions

   This meeting was organized to review scientific progress in
research on recombinant DNA molecules and to discuss appropriate
ways to deal with the potential biohazards of this work. Impressive scientific achievements have already been made in this field,
and these techniques have a remarkable potential for furthering
our understanding of fundamental biochemical processes in pro- and
eukaryotic cells. The use of recombinant DNA methodology promises
to revolutionize the practice of molecular biology. Although there
has as yet been no practical application of the new techniques,
there is every reason to believe that they will have significant
practical utility in the future.

   Of particular concern to the participants at the meeting was
the issue of whether the pause in certain aspects of research in
this area, called for by the Committee on Recombinant DNA Molecules of the National Academy of Sciences in the letter published
in July 1974 (*I*), should end, and, if so, how the scientific work
could be undertaken with minimal risks to workers in laboratories,
to the public at large, and to the animal and plant species shar-
ing our ecosystems.

   The new techniques, which permit combination of genetic
information from very different organisms, place us in an area of

biology with many unknowns. Even in the present, more limited
conduct of research in this field, the evaluation of potential
biohazards has proved to be extremely difficult. It is this
ignorance that has compelled us to conclude that it would be wise
to exercise considerable caution in performing this research.
Nevertheless, the participants at the Conference agreed that most
of the work on construction of recombinant DNA molecules should
proceed, provided that appropriate safeguards, principally biological and physical barriers adequate to contain the newly created
organisms, are employed. Moreover, the standards of protection
should be greater at the beginning and modified as improvements in
the methodology occur and assessments of the risks change. Furthermore, it was agreed that there are certain experiments in which
the potential risks are of such a serious nature that they ought
not to be done with presently available containment facilities.
In the longer term serious problems may arise in the large-scale
application of this methodology in industry, medicine, and agriculture. But it was also recognized that future research and
experience may show that many of the potential biohazards are less
serious and/or less probable than we now suspect.

II. Principles Guiding the Recom-
mendations and Conclusions

Although our assessments of the risks involved with each of
the various lines of research on recombinant DNA molecules may
differ, few, if any, believe that this methodology is free from
any risk. Reasonable principles for dealing with these potential
risks are: (i) that containment be made an essential consideration
in the experimental design and (ii) that the effectiveness of the
containment should match, as closely as possible, the estimated
risk. Consequently, whatever scale of risks is agreed upon, there
should be a commensurate scale of containment. Estimating the

risks will be difficult and intuitive at first, but this will
improve as we acquire additional knowledge; at each stage we shall
have to match the potential risk with an appropriate level of containment. Experiments requiring large-scale operations would seem
to be riskier than equivalent experiments done on a small scale
and therefore require more stringent containment procedures. The
use of cloning vehicles or vectors (plasmids, phages) and bacterial hosts with a restricted capacity to multiply outside of the
laboratory would reduce the potential biohazard of a particular
experiment. Thus, the ways in which potential biohazards and different levels of containment are matched may vary from time to
time, particularly as the containment technology is improved. The
means for assessing and balancing risks with appropriate levels of
containment will need to be reexamined from time to time. Hopefully, through formal and informal channels of information within
and between nations of the world, the way in which potential biohazards and levels of containment are matched would be consistent.

Containment of potentially biohazardous agents can be achieved
in several ways. The most significant contribution to limiting
the spread of the recombinant DNA's is the use of biological barriers. These barriers are of two types: (i) fastidious bacterial
hosts unable to survive in natural environments and (ii) nontransmissible and equally fastidious vectors (plasmids, bacteriophages, or other viruses) able to grow only in specified hosts.
Physical containment, exemplified by the use of suitable hoods or,
where applicable, limited access or negative pressure laboratories,
provides an additional factor of safety. Particularly important
is strict adherence to good microbiological practices which, to a
large measure, can limit the escape of organisms from the experimental situation and thereby increase the safety of the operation.
Consequently, education and training of all personnel involved in
the experiments is essential to the effectiveness of all containment measures. In practice, these different means of containment
will complement one another and documented substantial improvements

in the ability to restrict the growth of bacterial hosts and
vectors could permit modifications of the complementary physical
containment requirements.

Stringent physical containment and rigorous laboratory procedures can reduce but not eliminate the possibility of spreading
potentially hazardous agents. Therefore, investigators relying
upon "disarmed" hosts and vectors for additional safety must
rigorously test the effectiveness of these agents before accepting
their validity as biological barriers.

*III. Recommendations for Matching Types of
       Containment with Types of Experiments*

No classification of experiments as to risk and no set of containment procedures can anticipate all situations. Given our
present uncertainties about the hazards, the parameters proposed
here are broadly conceived and meant to provide provisional guidelines for investigators and agencies concerned with research on
recombinant DNA's. However, each investigator bears a responsibility for determining whether, in his particular case, special
circumstances warrant a higher level of containment than is suggested here.

A. Types of Containment.

1) Minimal risk. This type of containment is intended for
experiments in which the biohazards may be accurately assessed
and are expected to be minimal. Such containment can be achieved
by following the operating procedures recommended for clinical
microbiological laboratories. Essential features of such facilities are no drinking, eating, or smoking in the laboratory, wearing laboratory coats in the work area, the use of cotton-plugged

pipettes or preferably mechanical pipetting devices, and prompt disinfection of contaminated materials.

2) Low risk. This level of containment is appropriate for experiments which generate novel biotypes but where the available information indicates that the recombinant DNA cannot alter appreciably the ecological behavior of the recipient species, increase significantly its pathogenicity, or prevent effective treatment of any resulting infections. The key features of this containment (in addition to the minimal procedures mentioned above) are a prohibition of mouth pipetting, access limited to laboratory personnel, and the use of biological safety cabinets for procedures likely to produce aerosols (for example, blending and sonication). Though existing vectors may be used in conjunction with low-risk procedures, safer vectors and hosts should be adopted as they become available.

3) Moderate risk. Such containment facilities are intended for experiments in which there is a probability of generating an agent with a significant potential for pathogenicity or ecological disruption. The principal features of this level of containment, in addition to those of the two preceding classes, are that transfer operations should be carried out in biological safety cabinets (for example, laminar flow hoods), gloves should be worn during the handling of infectious materials, vacuum lines must be protected by filters, and negative pressure should be maintained in the limited access laboratories. Moreover, experiments posing a moderate risk must be done only with vectors and hosts that have an appreciably impaired capacity to multiply outside of the laboratory.

4) High risk. This level of containment is intended for experiments in which the potential for ecological disruption or pathogenicity of the modified organism could be severe and thereby

pose a serious biohazard to laboratory personnel or the public.
The main features of this type of facility, which was designed to
contain highly infectious microbiological agents, are its isolation from other areas by air locks, a negative pressure environment, a requirement for clothing changes and showers for entering
personnel, and laboratories fitted with treatment systems to
inactivate or remove biological agents that may be contaminants in
exhaust air and liquid and solid wastes. All persons occupying
these areas should wear protective laboratory clothing and shower
at each exit from the containment facility. The handling of agents
should be confined to biological safety cabinets in which the
exhaust air is incinerated or passed through Hepa filters. High-
risk containment includes, in addition to the physical and procedural features described above, the use of rigorously tested
vectors and hosts whose growth can be confined to the laboratory.

A. Types of Experiments. Accurate estimates of the risks
associated with different types of experiments are difficult to
obtain because of our ignorance of the probability that the anticipated dangers will manifest themselves. Nevertheless, experiments
involving the construction and propagation of recombinant DNA
molecules using DNA's from (*i*) prokaryotes, bacteriophages, and
other plasmids; (*ii*) animal viruses; and (*iii*) eukaryotes have
been characterized as minimal, low, moderate, and high risks to
guide investigators in their choice of the appropriate containment. These designations should be viewed as interim assignments
which will need to be revised upward or downward in the light of
future experience.

The recombinant DNA molecules themselves, as distinct from
cells carrying them, may be infectious to bacteria or higher organisms. DNA preparations from these experiments, particularly in
large quantities, should be chemically inactivated before disposal.

1) *Prokaryotes, bacteriophages, and bacterial plasmids*. Where the construction of recombinant DNA molecules and their propagation involve prokaryotic agents that are known to exchange genetic information naturally, the experiments can be performed in minimal-risk containment facilities. Where such experiments pose a potential hazard, more stringent containment may be warranted.

Experiments involving the creation and propagation of recombinant DNA molecules from DNA's of species that ordinarily do not exchange genetic information generate novel biotypes. Because such experiments may pose biohazards greater than those associated with the original organisms, they should be performed, at least, in low-risk containment facilities. If the experiments involve either pathogenic organisms or genetic determinants that may increase the pathogenicity of the recipient species, or if the transferred DNA can confer upon the recipient organisms new metabolic activities not native to these species and thereby modify its relationship with the environment, then moderate- or high-risk containment should be used.

Experiments extending the range of resistance of established human pathogens to therapeutically useful antibiotics or disinfectants should be undertaken only under moderate- or high-risk containment, depending upon the virulence of the organism involved.

2) *Animal viruses*. Experiments involving linkage of viral genomes or genome segments to prokaryotic vectors and their propagation in prokaryotic cells should be performed only with vector-host systems having demonstrably restricted growth capabilities outside the laboratory and with moderate-risk containment facilities. Rigorously purified and characterized segments of non-oncogenic viral genomes or of the demonstable nontransforming regions of oncogenic viral CNA's can be attached to presently existing vectors and propagated in moderate-risk containment facilities; as safer vector-host systems become available such experiments may be performed in low-risk facilities.

Experiments designed to introduce or propagate DNA from non-viral or other low-risk agents in animal cells should use only low-risk animal DNA's as vectors (for example, viral or mitochondrial), and manipulations should be confined to moderate-risk containment facilities.

3) *Eukaryotes.* The risks associated with joining random fragments of eukaryote DNA to prokaryotic DNA vectors and the propagation of these recombinant DNA's in prokaryotic hosts are the most difficult to assess.

A priori, the DNA from warm-blooded vertebrates is more likely to contain cryptic viral genomes potentially pathogenic for man than is the DNA from other eukaryotes. Consequently, attempts to clone segments of DNA from such animals and particularly primate genomes should be performed only with vector-host systems having demonstrably restricted growth capabilities outside the laboratory and in a moderate-risk containment facility. Until cloned segments of warm blood vertebrate DNA are completely characterized, they should continue to be maintained in the most restricted vector-host system in moderate-risk containment laboratories; when such cloned segments are characterized, they may be propagated as suggested above for purified segments of virus genomes.

Unless the organism makes a product known to be dangerous (for example, a toxin or virus), recombinant DNA's from cold-blooded vertebrates and all other lower eukaryotes can be constructed and propagated with the safest vector-host system available in low-risk containment facilities.

Purified DNA from any source that performs known functions and can be judged to be nontoxic may be cloned with currently available vectors in low-risk containment facilities. (Toxic here includes potentially oncogenic products or substances that might perturb normal metabolism if produced in an animal or plant by a resident microorganism.)

4) *Experiments to be deferred.* There are feasible experiments which present such serious dangers that their performance should

not be undertaken at this time with the currently available vector-
host systems and the presently available containment capability.
These include the cloning of recombinant DNA's derived from highly
pathogenic organisms (that is, Class III, IV, and V etiologic
agents as classified by the U.S. Department of Health, Education
and Welfare). DNA containing toxin genes, and large-scale experiments (more than 10 liters of culture) using recombinant DNA's
that are able to make products potentially harmful to man, animals,
or plants.

IV. Implementation

In many countries steps are already being taken by national
bodies to formulate codes of practice for the conduct of experiments with known or potential biohazard (2). Until these are
established, we urge individual scientists to use the proposals in
this document as a guide. In addition, there are some recommendations which could be immediately and directly implemented by the
scientific community.

A. Development of Safer Vectors and Hosts. An important and
encouraging accomplishment of the meeting was the realization that
special bacteria and vectors, which have a restricted capacity to
multiply outside the laboratory, can be constructed genetically,
and that the use of these organisms could enhance the safety of
recombinant DNA experiments by many orders of magnitude. Experiments along these lines are presently in progress and, in the
near future, variants of λ bacteriophage, nontransmissible plasmids, and special strains of *Escherichia coli* will become available. All of these vectors could reduce the potential biohazards
by very large factors and improve the methodology as well. Other
vector-host systems, particularly modified strains of *Bacillus
subtilis* and their relevant bacteriophages and plasmids, may also

be useful for particular purposes. Quite possibly safe and suitable vectors may be found for eukaryotic hosts such as yeast and
readily cultured plant and animal cells. There is likely to be a
continuous development in this area, and the participants at the
meeting agreed that improved vector-host systems which reduce the
biohazards of recombinant DNA research will be made freely available to all interested investigators.

B. Laboratory Procedures. It is the clear responsibility of
the principal investigator to inform the staff of the laboratory
of the potential hazards of such experiments, before they are
initiated. Free and open discussion is necessary so that each
individual participating in the experiment fully understands the
nature of the experiment and any risk that might be involved. All
workers must be properly trained in the containment procedures
that are designed to control the hazard, including emergency
actions in the event of a hazard. It is also recommended that
appropriate health surveillance of all personnel, including sero-
logical monitoring, be conducted periodically.

C. Education and Reassessment. Research in this area will
develop very quickly, and the methods will be applied to many different biological problems. At any given time it is impossible to
foresee the entire range of all potential experiments and make
judgments on them. Therefore, it is essential to undertake a continuing reassessment of the problems in the light of new scientific knowledge. This could be achieved by a series of annual
workshops and meetings, some of which should be at the international level. There should also be courses to train individuals
in the relevant methods, since it is likely that the work will be
taken up by laboratories which may not have had extensive experience in this area. High priority should also be given to research
that could improve and evaluate the containment of effectiveness
of new and existing vector-host systems.

V. New Knowledge

This document represents our first assessment of the potential
biohazards in the light of current knowledge. However, little is
known about the survival of laboratory strains of bacteria and
bacteriophages in different ecological niches in the outside
world. Even less is known about whether recombinant DNA molecules
will enhance or depress the survival of their vectors and hosts in
nature. These questions are fundamental to the testing of any new
organism that may be constructed. Research in this area needs to
be undertaken and should be given high priority. In general, however, molecular biologists who may construct DNA recombinant molecules do not undertake these experiments and it will be necessary
to facilitate collaborative research between them and groups
skilled in the study of bacterial infection or ecological microbiology. Work should also be undertaken which would enable us to
monitor the escape or dissemination of cloning vehicles and their
hosts.

Nothing is known about the potential infectivity in higher
organisms of phages or bacteria containing segments of eukaryotic
DNA, and very little is known about the infectivity of the DNA
molecules themselves. Genetic transformation of bacteria does
occur in animals, suggesting that recombinant DNA molecules can
retain their biological potency in this environment. There are
many questions in this area, the answers to which are essential
for our assessment of the biohazards of experiments with recombinant DNA molecules. It will be necessary to ensure that this
work will be planned and carried out; and it will be particularly
important to have this information before large-scale applications
of the use of recombinant DNA molecules are attempted.

REFERENCES AND NOTES

1. P. Berg *et al.*, *Science 185*, 303 (1974).
2. Advisory Board for the Research Councils, "Report of the
    Working Party on the Experimental Manipulation of the Genetic
    Composition of Micro-Organisms, Presented to Parliament by the
    Secretary of State for Education and Science by Command of Her
    Majesty, January 1975" (Her Majesty's Stationery Office,
    London, 1975); National Institutes of Health Recombinant DNA
    Molecule Program Advisory Committee, Bethesda, Maryland.
3. The work of the committee was assisted by the National
    Academy of Sciences--National Research Council Staff: Artemis
    P. Simopoulos, Executive Secretary, Division of Medical
    Sciences, Assembly of Life Sciences; Elena O. Nightingale,
    Resident Fellow, Division of Medical Sciences, Assembly of
    Life Sciences. Supported by the National Institutes of Health
    (contract NO1-OD-52103) and the National Science Foundation
    (grant GBMS75-05293). Requests for reprints should be
    addressed to: Division of Medical Sciences, Assembly of Life
    Sciences, National Academy of Sciences, 2101 Constitution
    Avenue, NW, Washington, D.C. 20418.

CHARGAFF AND SIMRING

*The thoughtful, measured tone of the early stages of the
debate soon erupted into a more intense, all-out fight. The
appearance of the following two letters ended whatever chances
there might have been for a compromise settlement of the issues.
Chargaff's "Frankenstein" hit the front pages of many newspapers
throughout the country and Simring's charges of a loaded committee
were not easily dismissed. They appeared under the heading, "On
the Dangers of Genetic Meddling."*

June 4, 1976*

On the Dangers of Genetic Meddling

A bizarre problem is posed by recent attempts to make so-
called genetic engineering palatable to the public. Presumably
because they were asked to establish "guidelines," the National
Institutes of Health have permitted themselves to be dragged into
a controversy with which they should not have had anything to do.
Perhaps such a request should have been addressed to the Department
of Justice. But I doubt that they would have wanted to become
involved with second-degree molecular biology.

Although I do not think that a terrorist organiation ever
asked the Federal Bureau of Investigation to establish guidelines
on the proper conduct of bombing experiments, I do not doubt what
the answer would have been; namely, that they ought to refrain
from doing anything unlawful. This also applies to the case under
discussion; no smokescreen, neither P3 nor P4 containment

---

facilities, can absolve an experimenter from having injured a
fellow being. I set my hope in the cleaning women and the animal
attendants employed in laboratories playing games with "recombinant DNA"; in the law profession, which ought to recognize a golden
opportunity for biological malpractice suits; and in the juries
that dislike all forms of doctors.

In pursuing my quixotic undertaking--fighting windmills with
an M.D. degree--I shall start with the cardinal folly, namely, the
choice of *Escherichia coli* as the host. Permit me to quote from a
respected textbook of microbiology (1): "*E. coli* is referred to as
the 'colon bacillus' because it is the predominant facultative
species in the large bowel." In fact, we harbor several hundred
different varieties of this useful microorganism. It is responsible for few infections but probably for more scientific papers
than any other living organism. If our time feels called upon to
create new forms of living cells--forms that the world has presumably not seen since its onset--why choose a microbe that has
cohabited, more or less happily, with us for a very long time
indeed? The answer is that we know so much more about *E. coli*
than about anything else, including ourselves. But is this a
valid answer? Take your time, study diligently, and you will
eventually learn a great deal about organisms that cannot live in
men or animals. There is no hurry, there is no hurry whatever.

Here I shall be interrupted by many colleagues who assure me
that they cannot wait any longer, that they are in a tremendous
hurry to help suffering humanity. Without doubting the purity of
their motives, I must say that nobody has, to my knowledge, set
out clearly how he plans to go about curing everything from
alkaptonuria to Zenker's degeneration let alone replacing or
repairing our genes. But screams and empty promises fill the air.
"Don't you want cheap insulin? Would you not like to have cereals
get their nitrogen from the air? And how about green man photo-
synthesizing his nourishment: 10 minutes in the sun for breakfast,

30 minutes for lunch, and 1 hour for dinner?" Well, maybe Yes,
and maybe No.

If Dr. Frankenstein must go on producing his little biological
monsters--and I deny the urgency and even the compulsion--why pick
E. coli as the womb? This is a field where every experiment is a
"shotgun experiment," not only those so designated; and who knows
what is really being implanted into the DNA of the plasmids which
the bacillus will continue multiplying to the end of time? And it
will eventually get into human beings and animals despite all the
precautions of containment. What is inside will be outside. Here
I am given the assurance that the work will be done with enfeebled
lambda and with modified, defective *E. coli* strains that cannot
live in the intestine. But how about the exchange of genetic
material in the gut? How can we be sure what would happen once
the little beasts escaped from the laboratory? Let me quote once
more from the respected textbook (*1*): "Indeed the possibility cannot be dismissed that genetic recombinant in the intestinal tract
may even cause harmless enteric bacilli occasionally to become
virulent." I am thinking, however, of something much worse than
virulence. We are playing with hotter fires.

It is not surprising, but it is regrettable that the groups
that entrusted themselves with the formulation of "guidelines,"
as well as the several advisory committees, consisted exclusively,
or almost exclusively, of advocates of this form of genetic
experimentation. What seems to have been disregarded completely
is that we are dealing here much more with an ethical problem than
with one in public health, and that the principal question to be
answered is whether we have the right to put an additional fearful
loan on generations that are not yet born. I use the adjective
"additional" in view of the unresolved and equally fearful problem
of the disposal of nuclear waste. Our time is cursed with the
necessity for feeble men, masquerading as experts, to make enor-
mously far-reaching decisions. Is there anything more far-reaching
than the creation of new forms of life?

Recognizing that the National Institutes of Health are not
equipped to deal with a dilemma of such import, I can only hope
against hope for congressional action. One could, for instance,
envision the following steps: (i) a complete prohibition of the
use of bacterial hosts that are indigenous to man; (ii) the creation of an authority, truly representative of the population of
this country, that would support and license research on less
objectionable hosts and procedures; (iii) all forms of "genetic
engineering" remaining a federal monopoly; (iv) all research
eventually being carried out in one place, such as Fort Detrick.
It is clear that a moratorium of some sort will have to procede
the erection of legal safeguards.

But beyond all this, there arises a general problem of the
greatest significance, namely, the awesome irreversibility of what
is being contemplated. You can stop splitting the atom; you can
stop visiting the moon; you can stop using aerosols; you may even
decide not to kill entire populations by the use of a few bombs.
But you cannot recall a new form of life. Once you have constructed a viable $E.$ $coli$ cell carrying a plasmid DNA into which a
piece of eukaryotic DNA has been spliced, it will survive you and
your children and your children's children. An irreversible
attack on the biosphere is something so unheard-of, so unthinkable
to previous generations, that I could only wish that mine had not
been guilty of it. The hybridization of Prometheus with Herostra-
tus is bound to give evil results.

Most of the experimental results published so far in this
field are actually quite unconvincing. We understand very little
about eukaryotic DNA. The significance of spacer regions, repeti-
tive sequences, and, for that matter, of heterochromatin is not
yet fully understood. It appears that the recombinant experiments in which a piece of animal DNA is incorporated into the DNA
of a microbial plasmid are being performed without a full appreci-
ation of what is going on. Is the position of one gene with
respect to its neighbors on the DNA chain accidental or do they

control and regulate each other? Can we be sure--to mention one
fantastic improbability--that the gene for a given protein hormone,
operative only in certain specialized cells, does not become car-
cinogenic when introduced naked into the intestine? Are we wise
in getting ready to mix up what nature has kept apart, namely the
genomes of eukaryotic and prokaryotic cells?

The worst is that we shall never know. Bacteria and viruses
have always formed a most effective biological underground. The
guerilla warfare through which they act on higher forms of life is
only imperfectly understood. By adding to this arsenal freakish
forms of life--prokaryotes propagating eukaryotic genes--we shall
be throwing a veil of uncertainties over the life of coming
generations. Have we the right to counteract, irreversibly, the
evolutionary wisdom of millions of years, in order to satisfy the
ambition and the curiosity of a few scientists?

This world is given to us on loan. We come and we go; and
after a time we leave earth and air and water to others who come
after us. My generation, or perhaps the one preceding mine, has
been the first to engage, under the leadership of the exact
sciences, in a destructive colonial warfare against nature. The
future will curse us for it.

Erwin Chargaff

350 Central Park West,
New York

REFERENCES

1. B. D. Davis, R. Dulbecco, H. N. Eisen, H. S. Ginsberg, W. B.
    Wood, Jr., *Microbiology* (Harper & Row, New York, ed. I, 1967),
    p. 769).

June 4, 1976*

The recombinant DNA research controversy is permeated by the
assumptions that (i) the work will go ahead; (ii) benefits outweigh the risks; (iii) we can act now and learn later; and (iv)
any given problem has a solution. We therefore had better take
those steps necessary to ensure that the April meeting of the
Recombinant DNA Molecule Program Advisory Committee at the
National Institutes of Health (NIH) is *not* the "last look before
the leap" (News and Comment, 16 Apr., p. 236).

Several serious questions must be addressed.

On what basis were the scientists on that committee chosen so
that a mutually reinforcing group was able to vote down almost
every safety suggestion requested by NIH director Donald S.
Frederickson? Why was there no committee discussion about or
reference to the myriad reports, statements, letters, and varied
data submitted to the committee from throughout the United States
by eminent scientists stressing the necessity for (i) more stringent control measures; (ii) centralized P4 facilities; (iii) rejection of *Escherichia coli* as host; or (iv) postponement of recombinant DNA research? When and why was it decided that the work will
go ahead merely pending guideline ratification?

As recently as this past February, after NIH's public hearings
on recombinant DNA research, David L. Bazelon, Chief Judge of the
District of Columbia Court of Appeals, advised that Frederickson,
in assessing the varied testimonies covering a spectrum ranging
from laboratory safety procedures to ramifications of interference
with evolution, should set forth in great detail the reason for
each step he takes or does not take. Yet at the April meeting,
the advisory committee reviewed the details of laboratory containment facilities and procedures as if the public hearings had

*Reprinted with permission from *Science, 192,* 940. Copyright
1976 by the American Association for the Advancement of Science.

never taken place. That the benefits outweigh the risks of recombinant DNA research was taken as a matter of course, not a matter
of discussion. This "act now and learn later" approach gave rise
to a vote for the use of an "enfeebled" strain of bacteriophage
(lambda), predicated in the proposed guidelines on its use with
"enfeebled" *E. coli* bacteria. Yet we cannot predict whether within the human organism, either host or vector, or both, will not
later revert to greater strength.

   We should seriously question whether these DNA committee
meetings are window dressing for those scientists, many currently
involved in recombinant DNA research, who are committed to pushing
this research ahead with as little impediment as possible. Producing guidelines serves not only as a sop to Cerberus but distracts
from the basic alternatives of (*i*) postponement of research and
(*ii*) open, unbiased discussions of benefits and risks. We should
be wary of self-imposed guidelines which experimenters may cite as
a defense in lawsuits for punitive monetary damages. It is none
too soon to consider federal legislation that would prevent limiting the liability of the experimenter, laboratory, institution,
manufacturer, distributor, and direct agent in the case of disease,
injury, or death resulting from recombinant DNA research.

   There are striking parallels between the recombinant DNA and
the nuclear energy controversies. Thirty years ago, when nuclear
energy development was initiated, the problems of waste transport
and disposal, sabotage, weapons proliferation, and low-level radiation were either not foreseen or not deemed worthy of consideration.

   Proponents of nuclear energy defined the problems and proposed
their own solutions. Questionable data were classified and talk
centered on design criteria, reactor safety, and regulation. The
unquantifiable problems—the genetic risk to future generations,
human fallibility, the vulnerability of centralized electric
generation, acts of malevolence, the threat to civil liberties by

massive security measures, and the economic investment and subsidies required--were not addressed.

In like manner, proponents of DNA research have set up the question of laboratory containment as the pivotal problem, for which their guidelines will be the solution. What scientist would claim that complete laboratory containment is possible and that accident due to human fallibility and technical failures will not inevitably occur?

It is therefore essential that open discussion include the entire range of problems in the field of genetic engineering and take into account the biohazards of accidental release of uncontrollable new organisms, the implications of interference with evolution, reduction of diversity in the gene pool, the imposition of complex medical decisions on individuals and society, and the inherent fallibility (not to mention corruptibility) of inspection, enforcement, and regulatory bodies.

We have the unique opportunity now, *before* the intellectual and economic investment in the development of recombinant DNA research grows much greater, to assess the benefits and risks. Such assessment should include acknowledgement that not all problems necessarily have solutions and that problems will arise that cannot possibly be foreseen. The vast number of human and technical variables precludes adequate anticipation of the problems of new technologies. At a recent recombinant DNA conference, NIH deputy director for science DeWitt Stetten, Jr., warned that ". . . the real hazard is the one no one around this table has dreamed of yet, and this you cannot specify against."

Francine Robinson Simring

Committee for Genetics
Friends of the Earth
72 Jane Street, New York

SINGER AND BERG

The joint responses by Singer and Berg, to these letters is
important not only for the arguments, but also for the fundamental
sense of honest betrayal and frustration conveyed. These appeared
under the heading, "Recombinant DNA: NIH Guidelines."

July 16, 1976*

Recombinant DNA: NIH Guidelines

Erwin Chargaff's and Francine R. Simring's letters (4 June,
pp. 938 and 940) regarding recombinant DNA research require
comment. Analysis of the history leading to, and the substance
of, the guidelines for conducting this line of research (*1*) suggest that both of these critics overlooked important facts.

It is relevant to our comments that we were among those who
first publicly expressed concern over the potential hazards of
recombinant DNA experiments (*2, 3*); we were members of the organi-
zing committee of the Asilomar Conference (*4*); neither of us is a
member of the National Institutes of Health (NIH) Program Advisory
Committee on Recombinant DNA, although we have been active commentators on that committee's efforts to develop guidelines; and one
of us is, and one is not, pursuing recombinant DNA experiments in
our own laboratories.

Chargaff questions the propriety and legitimacy of NIH's role
in formulating guidelines for recombinant DNA research. Certainly
the principal biomedical research arm of the United States must be
concerned with the health of laboratory workers and the public at
large. Even if Congress or another governmental agency had intervened early and assumed responsibility in the area of recombinant

---

*Reprinted with permission from *Science, 193*, 186-8. Copyright
1976 by the American Association for the Advancement of Science.

DNA research, it is not conceivable that policy could properly be formulated without the involvement of NIH and informed members of the scientific community. Acceptance of responsibility in this matter by the past and present directors of NIH was courageous, farseeing, and proper; moreover, the directors and the consultants who labored diligently to produce the guidelines deserve our gratitude.

Contrary to the implications in the letters by Chargaff and Simring, the discussions leading to the guidelines were directed toward eliminating or minimizing real and imagined hazards, rather than balancing benefits and risks. The only certain benefit is increased knowledge of basic biologic processes; the predicted benefits for medicine, agriculture, and industry will follow only upon this increased knowledge. It was concern for the potential risks with recombinant DNA that led a group of scientists involved in this research to call for a voluntary deferral of certain experiments (3). The guidelines either proscribe such experiments or require extremely stringent containment measures for them. Indeed, the list of experiments in the proscribed category was extended between 1974 (3) and the Asilomar Conference report (4) and is even further enlarged in the guidelines (1).

Permissible experiments under the guidelines are classified according to the best available estimate of potential risk. Increasing potential risk requires increasingly stringent biological and physical containment measures. Not all recombinant DNA experiments yield "new" organisms; recombination between the DNA's of organisms known to exchange genetic information in nature do not add uniquely man-made species to the biosphere. In these cases, the guidelines follow the general principle that the experiments are to be carried out under previously defined conditions for handling the most hazardous parent of the recombinant. When DNA from species that are not known to exchange genetic material in nature are recombined, additional precautions are required. And it is precisely concerns of the kind raised by Chargaff (for

example, the unpredictable consequences of intestinal colonization
by organisms carrying potentially harmful genes) that led to such
special precautions. Admittedly, the estimates of potential
hazard are presently conjectural and controversial. But it is
precisely for this reason that the requirements specified in the
guidelines are more stringent than most scientists estimate are
required for safety.

The adequacy of the containment prescribed for permissible
experiments is then the central issue. Estimates of achievable
containment levels must be based on available facts. We share
with Chargaff the belief that it is unacceptable to harm others.
But the recommended procedures are not "smokescreens." The P3 and
P4 levels of physical containment are designed specifically as
controls on accidental dispersal and human errors; they are
defined in detail in the guidelines, and there is documented
experience on which to judge the efficacy of these facilities.

With regard to biological containment, the encompassing
description of *Escherichia coli* as the "predominant facultative
species in the large bowel" in Chargaff's letter is misleading.
The guidelines require the use of strain K12 of *E. coli*. K12 rarely
establishes itself as a viable resident in the human gut (*1*). The
disabled derivatives of K12 must pass strict tests to establish
that they are unable to live in natural environments. Only thereafter can they be certified by the NIH Advisory Committee for use
in experiments requiring high levels of containment. The reports
from the April 1976 meeting of the Advisory Committee suggest that
the committee will be very cautious in its evaluation and certification of such systems. And, contrary to Chargaff's and Simring's statements, the problem of secondary natural recombination
of foreign sequences, either out of the original host and into
common enteric organisms, or between the experimental vectors and
naturally occurring organisms and vectors, has been central to the
discussions leading to definitions of disabled host-vector systems. In fact, the guidelines themselves describe and deal with

those problems. They require that data regarding the chance of
spread of the foreign DNA (by survival of a host cell or secondary recombination) in particular environments must be supplied;
the probability of such spread must be less than 10^{-8} before
certain experiments are permitted. Probabilities on the order of
10^{-8} afford a high level of confidence for achieving meaningful
containment, considering the small numbers of organisms that could
possibly escape as a result of human errors or flaws in physical
containment. Thirty years of study of the genetic chemistry of *E.
coli* K12 provides confidence that such levels of containment can
be achieved. While it is important to investigate alternatives
to *E. coli* K12, it is not at all certain that useful and safer
organisms exist. Predictions about the existence of rare and
fastidious organisms unable to exchange DNA with common organisms
inhabiting man or other living things are highly speculative.

If Chargaff and Simring had examined the massive and readily
available correspondence, minutes, and documents accumulated over
the last 3 years, they would have recognized that precisely those
matters they claim were disregarded were discussed in considerable
detail. (This documentation has been collected by the Massachusetts Institute of Technology Program in Oral History of Sciences.)
The charge that discussions have been "permeated by the assumption
that the work will go ahead" or that we can "act now and learn
later" is inconsistent with the 1974 moratorium and with the
acceptance at Asilomar, and in the guidelines, of the principle
that certain experiments should be deferred. The essence of the
development of the guidelines by the NIH Advisory Committee was a
discussion of alternative containment specifications, including
those mentioned by Simring. Simring fails to point out that the
letters to the director of NIH from eminent scientists and laymen
ran the full gamut in their views concerning the necessity for
more or less stringent control measures. Those same issues as
well as others raised by the public hearings held in February 1976
were subsequently reviewed in extraordinary depth by the director

and his staff. The director then asked the Advisory Committee to
again address certain issues. In April, after long and searching
consideration, the Advisory Committee reaffirmed certain earlier
recommendations and changed others. [This is discussed in the
lengthy commentary accompanying the guidelines (5).]

Chargaff and Simring urge a slow approach to experimentation.
It should be recognized that a "slow approach" is what was
achieved by the voluntary deferral and the Asilomar guidelines.
Research on recombinant DNA will proceed at only a fraction of the
possible rate because of the need for certified host-vector systems, acquisition of sophisticated physical containment facilities,
and the required deferral of a large group of interesting and
important experiments. Presently, in addition to a slow-down,
there is a far-reaching awareness on the part of investigators of
the need for caution, and a largely cooperative atmosphere exists
regarding the need for control of this type of experimentation.

Simring's attempt to draw analogies between recombinant DNA
and the nuclear energy controversies obscures the facts. The
discussions on recombinant DNA have been public since their beginning. The matter has been widely reported by the public press.
The publicity permitted all concerned individuals and groups to
enter the deliberations. No datum has been classified and no commentary has been withheld from the public. Indeed, most policy
has been developed in public sessions. In addition to containment,
the unquantifiable problems have been addressed. The problems may
be difficult, but they can be dealt with in a rational manner.

Finally, we are deeply disturbed by the distortions, derision,
and pessimism that permeate Chargaff's comments. He appears to
see science as a curse on our time, and men as feeble. In our
view it is knowledge and understanding derived from science and
scholarship that lead men to rationality and wisdom.

Maxine F. Singer

National Canter Institute
National Institutes of Health
Bethesda, Maryland

Paul Berg

Department of Biochemistry
Stanford University Medical Center
Stanford, California

REFERENCES

1. National Institutes of Health, *Guidelines for Research Involving Recombinant DNA Molecules* (National Institutes of Health, Washington, D.C., 1976).
2. M. Singer and D. Soll, *Science 181*, 1114 (1973).
3. P. Berg *et al.*, *ibid.*, *185*, 303 (1974).
4. P. Berg, D. Baltimore, S. Brenner, R. O. Roblin, M. F. Singer, *ibid.*, *188*, 991 (1975).
5. D. S. Fredrickson, *Decision of the Director, National Institutes of Health, to Release Guidelines for Research on Recombinant DNA Molecules* (National Institutes of Health, Washington, D.C., 1976).

Annotated Bibliography

John Richards[1]
Richard Michod[2]
John Davis Messer[3]

This bibliography lists articles, books and documents on the social, political and ethical aspects of recombinant DNA research, and the controversies arising from this research. We have omitted works devoted mainly to topics in human genetic engineering and genetic screening and counseling.

There are seven separate sections: I. books, II. government documents, III. journal articles and letters, IV. selected news coverage, V. feature articles in popular magazines and trade journals, VI. technical papers, and VII. a final section suggesting further readings. We have selected among the letters which have been published, and for news coverage we have relied basically on Nicholas Wade's extensive accounts in *Science*. Other news coverage is included only where it offers additional insights or more detailed accounts.

In cross-referencing, the Roman numerals refer to the section of the bibliography. For example, *Rogers (1977) I* refers to Michael Rogers's book, *Biohazard* listed in section I.

[1]Department of Philosophy, University of Georgia, Athens, Ga.

[2]Department of Ecology and Evolutionary Biology, University of Arizona, Tucson, Arizona.

[3]Department of Philosophy, University of Kentucky, Lexington, Kentucky.

337

I. BOOKS AND BOOK LENGTH MANUSCRIPTS

Beers, Roland F., and Bassett, Edward G., eds.   1977. *Recombinant Molecules: Impact on Science and Society*. Miles International Symposium Series, No. 10. New York: Raven Press.

> Technical volume, with the exception of section F: "Societal Impact--Issues and Policies."

Cooke, Robert. 1977. *Improving on Nature: The Brave New World of Genetic Engineering*. New York: Quadrangle/The New York Times Book Company.

Daedalus. 1978. *Limits of Scientific Inquiry*. Spring.

> As a result of the current interest in rDNA and in experimentation on human subjects, this entire issue of the journal is devoted to the question of limiting scientific inquiry.

Federation of American Scientists. 1976. *Public Interest Report*, 29 (April). Excerpted as "Issues in Recombinant DNA Research." In *Hospital Practice*, July, 1976, p. 20.

Freter, Rolf (to appear). *Real and Imagined Dangers of Recombinant DNA Technology: The Need for Expert Evaluation*. Ann Arbor: University of Michigan Press.

Goodfield, June. 1977. *Playing God: Genetic Engineering and the Manipulation of Life*. New York: Random House.

> A good account of the history of the rDNA debate which focuses on the breakdown of the existing social contract between science and society, and the new issues which need to be confronted. See also Goodfield (1977a) VII, and reviews: Medawar (1977) III, Lewin (1977), III, Goodell (1977) III, and Oppenheimer (1978) III.

Jackson, David A., and Stich, Stephen P., eds. (to appear). *The Recombinant DNA Debate*. Ann Arbor: University of Michigan Press.

> Good interdisciplinary collection, including contributions from Chakrabarty, Curtiss, Sinsheimer, Wald, Freter, Dworkin, Michael and Cohen.

Lappé, Marc, and Morison, Robert S., eds.   1976. *Ethical and Scientific Issues Posed by Human Uses of Molecular Genetics*. *Annals of the New York Academy of Sciences*, 265. New York: The New York Academy of Sciences.

See especially: "Reflections on Issues Posed by Recombinant DNA Molecule Technology, Pt. I Richard Roblin, 59-65; Pt. II Stanfield Rogers, 66-70; and Pt. III Alexander Morgan Capron, 71-81.

Packard, Vance. 1977. *The People Shapers*. Massachusetts: Little Brown.

rDNA is discussed only briefly, but see review: Oppenheimer (1978) III.

Research With Recombinant DNA. NAS 1977. Washington, D.C.: National Academy of Sciences.

Extensive collection of papers delivered at the Academy Forum, March 7-9, 1977.

Rifkin, Jeremy, and Howard, Ted. 1977. *Who Should Play God?* New York: Delacorte Press.

Rogers, Michael. 1977. *Biohazard*. New York: Alfred A. Knopf.

Good personal account of the early developments from Asilomar through the NIH June 1976 Guidelines by the science writer for *Rolling Stone*. See reviews: Medawar (1977) III, Lewin (1977) III, Goodell (1977) III, Oppenheimer (1978) III, and Beckwith (1978) III.

Saul, George B., II. 1976. *Genetic Engineering: Source of Hope and Concern*. Vermont: Middlebury College.

Lecture delivered at Middlebury College, February 17, 1976.

Science, 196. April 8.

Entire issue devoted to rDNA research. Mostly technical reports with the exception of Singer (1977) III, and J. Abelson (1977) III.

Scott, W. A., and Werner, R. A., eds. 1977. *Molecular Cloning of Recombinant DNA*. New York: Academic Press.

Proceedings of the Miami Winter Symposia, January 1977. Technical with the exception of Berg, and Curtiss *et al*.

Stencel, Sandra. 1977. "Genetic Research." *Editorial Research Reports,* March, 225-244.

Wade, Nicholas. 1977. *The Ultimate Experiment: Man-Made Evolution*. New York: Walker and Company.

The "News and Comment" writer for *Science* integrates
his many articles on rDNA into a detailed and accurate
account of the development of the controversy. See
reviews: Goodell (1977) III, Lewin (1977) III,
Medawar (1977) III, and Beckwith (1978) III.

Winkel, Gary; Glass, David C.; and Fulco, Adrienne, eds. (to appear).
Social and Ethical Implications of Science and Technology.
New York: Plenum Press.

II. GOVERNMENT DOCUMENTS AND REPORTS

California. Assembly. Committee on Health. Committee on
Resources, Land Use, and Energy. Joint Hearings. Wagstaff,
Kenneth J. 1977. *Background Paper for Joint Hearings on the
Health and Environmental Implications of Recombinant DNA
Research*. Sacramento: Assembly Health Committee.

Cambridge, Massachusetts. The Cambridge Experimentation Review
Board. CERB 1976. *Report,* August 6, 1976. Reprinted in
Bulletin of the Atomic Scientists, 33 (May), 23–27.

Great Britain. Parliament. Advisory Board for the Research
Councils (Ashby Report). *Report of the Working Party on the
Experimental Manipulation of the Genetic Composition of Micro-
Organisms* (1975). London: Her Majesty's Stationery Office;
January 21, Cmnd. 5880.

_____. *Report of the Working Party on the Practice of Genetic
Manipulation* (Williams Report). London: Her Majesty's Stationery Office, August 1976. Cmnd 6600.

New York State. Feinberg, Deborah W., and Berger, Richard G.
1977. *Report and Recommendations of the New York State
Attorney General on Recombinant DNA Research*. New York:
Office of the Attorney General.

U.S. Congress. House. Committee on Interstate and Foreign Com-
merce. Subcommittee on Health and Environment. *Recombinant
DNA Research Act of 1977* (Rogers Hearings, 1977). Washington,
D.C.: U.S. Government Printing Office.

March 15–17 hearings. A comprehensive collection containing extensive testimony.

_____. Committee on Science and Astronautics. Subcommittee on Science, Research and Development. *Genetic Engineering: Evolution of a Technological Issue* (Davis report, 1974). Washington, D.C.: U.S. Government Printing Office.

This report, by James M. McCullough, is an update of a 1972 report by the same name which attempts to describe the developing controversy over rDNA.

_____. Senate. Committee on Human Resources. *Recombinant DNA Safety Regulation Act* (Kennedy Report, 1977). Washington, D.C.: U.S. Government Printing Office.

Report submitted by Senator Kennedy July, 1977, to accompany S. 1217. Includes a summary of the legislation and of Kennedy Hearings 1975, 1976, and 1977, along with a detailed analysis of the legislation.

_____. Subcommittee on Health and Scientific Research. Powledge, Tabitha M., and Dach, Leslie. 1977. *Biomedical Research and the Public*. Washington, D.C.: U.S. Government Printing Office.

Proceedings of the Airlie House conference, Warrenton, Virginia, April 1-3, 1976.

_____. *Recombinant DNA Regulation Act, 1977*. (Kennedy Hearings 1977). Washington, D.C.: U.S. Government Printing Office.

April 6, 1977 hearings, "to regulate activities involving rDNA."

_____. Committee on Labor and Public Welfare. Subcommittee on Health. *Genetic Engineering, 1975*. (Kennedy Hearings, 1975). Washington, D.C.: U.S. Government Printing Office.

April 22, 1975 hearings, "Examination of the relationship of a free society and its scientific community." Includes testimony by S. N. Cohen, D. Brown, W. Gaylin and H. Holman.

_____. Committee on Labor and Public Welfare. Subcommittee on Health. Committee on the Judiciary. Subcommittee on Administrative Practice and Procedure. Joint Hearing *Oversight Hearing on Implementation of NIH Guidelines Governing Recombinant DNA Research*. (Kennedy Hearings 1976). Washington, D.C.: U.S. Government Printing Office.

September 22, 1976 hearings: "Examination of NIH Guidelines governing rDNA research." Includes testimony by D. Fredrickson, D. Baltimore, B. Zimmerman, and C. J. Stetler.

U.S. Department of Health, Education and Welfare. Federal Interagency Committee on Recombinant DNA Research. *Interim Report:
   Suggested Elements for Legislation* (Interagency 1977). Submitted to the Secretary, H.E.W. March 15, 1977.

_____. Public Health Service. National Institutes of Health.
   *Recombinant DNA Research Volume I: Documents Relating to "NIH
   Guidelines for Research Involving Recombinant DNA Molecules"
   February 1975-June 1976.* (NIH Documents 1976). Washington,
   D.C.: U.S. Government Printing Office.

      Contains: July 1976 NIH Guidelines; Asilomar Guidelines;
      January 1976 NIH Guidelines; Proceedings of a Conference;
      letters on the Guidelines; and additional reports and
      general information. Extremely valuable collection.

_____. *Recombinant DNA Research Guidelines: Draft Environmental
   Impact Statement.* (NIH, 1976a). *Federal Register,* September
   9, 1976.

_____. *Recombinant DNA Research: Proposed Revised Guidelines.*
   (NIH, 1977). *Federal Register,* September 27, 1977.

_____. *Final Environmental Impact Statement on NIH Guidelines
   for Research Involving Recombinant DNA Molecules,* two volumes.
   (NIH 1977a). October. Washington, D.C.: U.S. Government
   Printing Office.

III. JOURNAL ARTICLES

Abelson, John. 1977. "Recombinant DNA: Examples of Present-Day
   Research." *Science, 196* (April), 159-160.

      Introductory survey piece for this issue of *Science,*
      which is devoted entirely to rDNA.

Abelson, Phillip H. 1977. "Recombinant DNA." *Science, 197*
   (August), 721.

      Editorial.

Bayev, A. A. 1974. *Gene Engineering: Reality and Promise.*
   Arlington, Va.: U.S. Joint Publications Research Service. Trans-
   lated from *Nauka I Zhizn'* (Russian), 5 (May), 18-26.

      Early account of the development of the technology.

Becker, Frank. 1977. "Law vs. Science: Legal Control of Genetic
   Research." *Kentucky Law Journal, 65,* 880-894.

   Public input into scientific decision-making is necessary,
   but must not become absolute control. The right to regulate rDNA must be tempered by a respect for the rights of
   the scientist.

Beckwith, Jon. 1978. "Recombinant DNA--The End of the Beginning
   of a Controversy." *Trends in Biological Science, 3* (April),
   94-5.

   Review of Rogers 1977 I, and Wade 1977 I.

Berg, Paul. 1976. "Genetic Engineering: Challenge and Responsibility." *ASM News, 42,* 273-277.

_____. 1977. Recombinant DNA Research Can be Safe." *Trends
in Biological Science, 2* (February). N25-27.

   The NIH Guidelines "afford the security needed to meet the
   perceived risks." Governing bodies must encourage scientists' participation, and must establish means to channel
   their input into the determination of policy.

Berg, Paul. 1976. "Genetic Engineering: Challenge and Responsibility." *ASM News, 42,* 273-277.

Berg, Paul, *et al.* 1974. (Letter) "Potential Biohazards of
   Recombinant DNA Molecules." *Science, 185* (July), 303.
   Appeared simultaneously in *Proceedings of the National Academy
   of Sciences USA, 71,* 2593-2594. *Reprinted in this volume:
   Appendix.*

   Report of the Berg committee.

Berg, Paul, *et al.* 1975. "Asilomar Conference on Recombinant DNA
   Molecules." *Science, 188* (June), 991-994. *Reprinted in this
   volume: Appendix.*

   Summary statement of the report submitted by the executive
   committee of the conference.

Callahan, Daniel. 1977. "Recombinant DNA: Science and the
   Public." *Hastings Center Report, 7* (April), 20-23.

   A consideration of the role of the public in the debate.
   What options are open, and what ethical and social
   criteria are appropriate?

Chakrabarty, Anando M. 1976. "Which Way Genetic Engineering?"
   Industrial Research, 18 (January), 45-50.

   Applications in industry.

Chargaff, Erwin. 1975. "Profitable Wonders." *The Sciences, 15* (August/September), 21-26. Excerpt, "A Slap at the Bishops of Asilomar." *Science, 190* (October), 135.

_____. 1976. (Letter) "On the Dangers of Genetic Meddling." *Science, 192* (June), 938-940. *Reprinted in this volume: Appendix.*

Cohen, Stanley N. 1975. "The Manipulation of Genes." *Scientific American, 233* (July), 24-33.

>   Excellent introduction and review of the technical and historical developments leading to recombinant DNA techniques.

_____. 1977. "Recombinant DNA: Fact and Fiction." *Science, 195* (February), 654-657.

>   The NIH guidelines provide an added measure of safety. There is no interference with "evolutionary wisdom," and the freedom of scientific inquiry is a bogus issue. This is a powerful paper, and provides one of the best defenses of continuing research under the Guidelines.

Curtiss, Roy, III. 1976. "Genetic Manipulation of Microorganisms: Potential Benefits and Biohazards." *Annual Review of Microbiology, 30,* 507-533.

>   Curtiss's lab has genetically engineered strains of *E. coli* to facilitate safer research. This is a detailed account of the plausibility of potential biohazards and the means to contend with them. See Curtiss *(this volume),* and also Durden-Smith (1977).

_____. 1978. "Biological Containment and Cloning Vector Transmissibility." *Journal of Infectious Diseases* (in press).

_____, *et al.* 1977. "Biohazard Assessment of Recombinant DNA Molecule Research." In Mitsuhashi, S.; Rosival, L.; and Krcmery, V. (Eds.) *Plasmids Medical and Theoretical Aspects.* Prague: Avicenum, Czechoslovak Medical Press, 375-387.

_____, *et al.* 1977. "Construction and Use of Safer Bacterial Host Strains for Recombinant DNA Research." In Scott and Werner (1977), 99-114.

>   See Maturin and Curtiss (1977), III.

Davis, Bernard D. 1974. "Genetic Engineering: How Great Is the Danger?" *Science, 186* (October), 309.

>   Editorial.

_____. 1976. (Letter) "Evolution, Epidemiology and Recombinant DNA." *Science, 193* (August), 442.

_____. 1977. "The Recombinant DNA Scenarios: Andromeda Strain, Chimera, and Golem." *American Scientist, 65* (September/October), 547-555.

An analysis from the perspective of epidemiology and evolutionary biology, concluding that the three scenarios in the title are a "product of man's literary imagination, and not his technology."

_____, and Sinsheimer, Robert L. 1976. "Discussion Forum: The Hazards of Recombinant DNA." *Trends in Biological Science, 1* (August), N178-N180.

Day, P. R. 1977. "Plant Genetics: Increasing Crop Yield." *Science, 197,* 1334-1339.

Survey article on plant genetics, including discussion of rDNA.

Dismukes, Key. 1977. "Recombinant DNA: A Proposal for Regulation." *Hastings Center Report, 7* (April), 25-30.

The NIH Guidelines do not provide the protection intended by the authors. Refinements are proposed. It is necessary to develop a broader, more systematic, policy which does not treat rDNA as an isolated case.

Eisinger, J. 1975. "The Ethics of Human Gene Manipulation." *Federation Proceedings, 34* (May), 1418-1420.

Introductory remarks to a symposium with a summary of the discussion. See also Roblin (1975) III, and Lappé (1975) III.

Frazier, Kendrick. 1975. "Rise to Responsibility at Asilomar." *Science News, 107* (March 22), 187.

Editorial.

Gardner, Barbara J. 1976. "The Potential for Genetic Engineering: A Proposal for International Legal Control." *Virginia Journal of International Law, 16,* 403-429.

Ad hoc mechanisms like the Asilomar conference must be replaced with an International Ethics Review Board.

Gaylin, Willard. 1977. "The Frankenstein Factor." *New England Journal of Medicine, 297,* 665-667.

"'Gene Engineering'--A Genie Out of the Test Tube" (LG 1975).
    Dangers of Gene Engineering Discussed, 1-9. Arlington, Va.:
    U.S. Joint Publications Research Service. Translated from
    Literaturnaya Gazeta (Russian), February 26, p. 13.

>   Condensed version of Berg (1974) III, and an interview
>   with O. Baroyan on the dangers of intentional alteration
>   of genes.

Goodell, Rae (1978). "Literature Guide: Review of Recent Books on
    the rDNA Controversy." *Newsletter on Science, Technology and
    Human Values, 22* (January), 25-29. Edited and revised version
    of "The Alchemy Man," *The Washington Post,* September 11, 1977.

>   Review of Goodfield (1977) I, Rogers (1977) I, and Wade
>   (1977) I.

Greenberg, Daniel S. 1977. "Lessons of the DNA Controversy."
    The New England Journal of Medicine, 297 (November), 1187-
    1188.

>   Argues against viewing the move to "go public" as a
>   mistake.

Grobstein, Clifford. 1976. "Recombinant DNA Research: Beyond the
    NIH Guidelines." *Science, 194* (December), 1133-1135.

>   Charges that the NIH committee was by its makeup bound
>   to ignore ecological hazards, and to avoid ethical,
>   social and political issues. The control of the research
>   should be taken out of the hands of NIH and placed in a
>   less specialized governmental organization, *e.g.*, a presi-
>   dential commission.

_____. 1977. "The Recombinant DNA Debate." *Scientific
    American, 237* (July), 22-33.

Halvorson, Harlyn O. 1977. "Recombinant DNA Legislation--What
    Next?" *Science, 198* (October), 357.

>   Editorial.

Hardin, Garrett J. 1975. "The Fateful Quandary of Genetic
    Research." *Prism, 3* (March), 20-23+.

Helinski, Donald R. 1978. "Plasmids as Vehicles for Gene Clon-
    ing: Impact on Basic and Applied Research." *Trends in Bio-
    logical Science, 3* (January), 10-14.

>   Describes procedures for using plasmid elements as
>   vehicles for cloning DNA from any source in *E. coli.*
>   Considers impact, and technical problems.

Helling, Robert B.   1975.   "Eukaryotic Genes in Prokaryotic Cells."
   Stadler Genetics Symposium, 7, 15–36.

_____, and Allen, Sally L.   1976.   "Freedom of Inquiry and
   Scientific Responsibility." *BioScience, 26,* 609–610.

   The issue of human genetic manipulation is not relevant
   to rDNA research, and neither is illicit use. The series
   of events leading to the survival of a pathogenic organism are extremely improbable, and the stringent NIH
   Guidelines are more than sufficient protection.

Holliday, Robin.   1977.   "Should Genetic Engineers be Contained?"
   New Scientist, 73 (February), 399–401.

   Detailed account of the probability of events in a
   hypothetical scenario of a man-made epidemic. Concludes
   that the chance of accident from rDNA research is
   "vanishingly small."

Hoskins, B. B., *et al.*   1977.   "Application of Genetic and
   Cellular Manipulations to Agricultural and Industrial Problems." *BioScience, 27* (March), 188–191.

   The earliest and most widespread application of cellular
   and genetic manipulation will be to industrial and
   agricultural processes. This details some of the bene-
   ficient possibilities "which have been largely ignored."

Hubbard, Ruth.   1976.   "Gazing into the Crystal Ball." *BioScience,
   26* (October), 608, 611.

   The NIH Guidelines are an exercise in Newspeak. Legislation and agencies are required at both the national and
   international level to rule out rDNA from different
   species, or restrict it to a few high containment facilities.

Jonas, Hans.   1976.   "Freedom of Scientific Enquiry and the Public
   Interest." *Hastings Center Report, 6* (August), 15–17.

_____.   1977.   "Science Involves Action," and "Ethics, Law
   Examine Experiments." Parts 11 and 12, "Moral Choices in
   Contemporary Society." *National Catholic Reporter,* April 15,
   pp. 16, 18.

   Much scientific research has crossed the boundary between
   contemplation and action, consequently freedom of inquiry
   no longer should be guaranteed. See also Jonas *(this
   volume).*

(Kassinger, Ted, and Solomon, Benna). 1977. "Recombinant DNA and Technology Assessment." *Georgia Law Review*, 11 (Summer), 785-878.

> Excellent account of the legal implications and responsibilities.

King, Jonathan. 1977. "A Science for the People." *New Scientist*, 74 (June), 634-636.

> Account of the development of Science for the People, and the events in Cambridge. Argues that the issue of rDNA can be seen as a vehicle for promoting the democratization of science.

Lappé, Marc. 1975. "The Human Use of Molecular Genetics." *Federation Proceedings*, 34 (May), 1425-27.

> Highlights the ethical problems implicit in setting goals and priorities for genetic techniques. See also Roblin (1975) III, and Eisinger (1975) III.

Lederberg, Joshua. 1975. "DNA Splicing: Will Fear Rob Us of Its Benefits?" *Prism*, 3 (November), 33-37.

Lefkowitz, Louis J. 1977. "A Legal Officer's Dilemma." *Bulletin of the Atomic Scientists*, 33 (May), 11.

Lewin, Roger. 1977a. "Hook Up in Genetic Engineering." *New Scientist*, 76 (November), 432, 434.

> Review of Goodfield (1977) I, and Rogers (1977) I.

Macklin, Ruth. 1977. "On the Ethics of *Not* Doing Scientific Research." *Hastings Center Report*, 7 (December), 11-13.

Maturin, L., Sr., and Curtiss, Roy, III. 1977. "Degradation of DNA by Nucleases in Intestinal Tracts of Rats." *Science*, 196, 216-218.

May, William F. 1978. "The Right to Know and the Right to Create." *Newsletter on Science, Technology & Human Values*, 23, 34-41.

Medawar, P. B. 1976. "The Scientific Conscience." *Hospital Practice*, 11 (July), 17, 20.

_____. 1977. "Fear and DNA." *New York Review of Books*, 24 (October), 15-20.

> Review of Goodfield (1977) I, Rogers (1977) I, and Wade (1977) I.

Michael, Donald N. 1977. "Who Decides Who Decides: Some Dilemmas and Other Hopes." In Stich and Jackson (to appear). An excerpt appears in *Hastings Center Report, 7* (April), 23.

Novick, Richard P. 1977. "Present Controls are Just a Start." *Bulletin of the Atomic Scientists, 33* (May), 16-22.

Oppenheimer, Jane M. 1978. "Biology and Society." *American Scientist , 66* (March/April), 223.

      Review of Goodfield (1977) I, Rogers (1977) I, and Packard (1977) I.

Ptashne, Mark. 1976. "The Defense Doesn't Rest." *The Sciences, 16* (September/October), 11-12.

      Response to Wald (1976) III.

"Recombinant DNA Debate Three Years On." (Nature 1977). *Nature, 268* (July), 185.

      Editorial.

Roblin, Richard. 1975. "Ethical and Social Aspects of Experimental Gene Manipulation." *Federation Proceedings, 34* (May), 1421-1424.

      Good discussion of the ethical paradoxes in the rDNA controversy.

Rowe, Wallace P. 1977. "Guidelines That Do the Job." *Bulletin of the Atomic Scientists, 33* (May), 14-15.

      _____. 1977a. "Recombinant DNA, What Happened?" *New England Journal of Medicine, 297*, 1176-1177.

      Editorial.

Shanmugam, K. T., and Valentine, Raymond C. 1975. "Molecular Biology of Nitrogen Fixation: Manipulation of Nitrogen Fixation Genes May Lead to Increased Production of High-Quality Protein." *Science, 187* (March), 919-924.

      _____; O'Gara, F.; and Valentine, R. C. 1978. "Genetic Engineering of New Nitrogen Fixing Plants." In Hollaender, Alexander, ed. *Report of the Public Meeting on Genetic Engineering for Nitrogen Fixation, October 5-6, 1977*. National Science Foundation. Washington, D.C.: U.S. Government Printing Office , 31-41.

Simring, Francine. 1976. (Letter) "On the Dangers of Genetic
    Meddling." *Science, 192* (June), 940. *Reprinted in this volume:
    Appendix.*

Singer, Maxine F. 1976. "Summary of the Proposed Guidelines."
    ASM News, 42, 277–287.

_____. 1977. "The Recombinant DNA Debate." *Science, 196,*
    127.

        Editorial.

_____. 1977a. "Scientists and the Control of Science." *New
    Scientist, 74* (June), 631–634.

        An analysis of the course of the debate, defending the
        decision to "go public," and arguing that it is essential
        to develop more effective means to inform and educate the
        public about science.

_____, and Berg, Paul. 1976. (Letter) "Recombinant DNA: NIH
    Guidelines." *Science, 193* (July), 186, 188. *Reprinted in
    this volume: Appendix.*

Singer, Maxine, and Soll, Dieter. 1973. (Letter) "Guidelines for
    DNA Hybrid Molecules." *Science, 181* (September), 1114.
    Reprinted in this volume: Appendix.

Sinsheimer, Robert L. 1969. "The Prospect for Designed Genetic
    Change." *American Scientist, 57,* 134–142.

        Early account of the potential for applying rDNA tech-
        nology to insulin production set within the broader
        context of human genetic engineering.

_____. 1975. "Troubled Dawn for Genetic Engineering." *New
    Scientist, 68* (October), 148–151.

        Survey of the new responsibilities and potentials of
        rDNA research. This is a good account of some of the
        broader moral issues.
_____. 1976. "Recombinant DNA––On Our Own." *BioScience, 26*
    (October), 599.

        Editorial.

_____. 1977. "An Evolutionary Perspective for Genetic
    Engineering." *New Scientist, 73* (January), 150–152.

        The NIH guidelines are too narrowly conceived, since they
        are based on a public health model and fail to take into
        account the implications of crossing natural evolutionary
        barriers. This argument has become one of the central
        issues in the debate. See Cohen (1977) III, for a response.

_____, and Piel, Gerard. 1976. "Inquiry into Inquiry: Two
Opposing Views." *Hastings Center Report*, *6* (August), 18-19.

> Excerpted from responses given to Jonas' remarks at the
> Airlie House Conference (see Powledge and Dach (1977) I).
> Sinsheimer argues that there are legitimate reasons for
> curbing both the means and ends of scientific research.
> Piel responds that a scientist can, and should, accept no
> authority in this but his own judgment and conscience.

Stettin, DeWitt, Jr. 1975. "Freedom of Inquiry." *Genetics*, *81*,
412-425.

_____. 1975a. "Freedom of Inquiry." *Science*, *189* (September),
953.

> Editorial.

Szabo, G. S. A. 1977. "Patents and Recombinant DNA." *Trends
in Biological Science*, 2 (November), N246-N249.

Thomas, Lewis. 1977. "Notes of a Biology-Watcher: The Hazards
of Science." *New England Journal of Medicine*, *296* (February),
324-328.

> There is no moral limit to scientific inquiry, or to
> knowledge. rDNA cannot serve as a test case for this
> since we already possess the knowledge. Also argues, as
> in (1974) VII, that it takes a long time for a microbe
> to become a pathogen so this is a groundless fear.

_____. 1978. "Hubris in Science?" *Science*, *200*, 1459-62.

Tooze, John. 1976. "Practical Guidelines." *Trends in Biological Science*, *1* (November), N246-7.

> Discussion of the Williams report.

_____. 1977. "Genetic Engineering in Europe." *New Scien-
tist*, *73* (March), 592-594.

Toulmin, Stephen. 1977. "DNA and the Public Interest." *New
York Times*, March 12, p. 23.

> Editorial.

_____. 1978. Review of Portugal, F. H., and Cohen, J. S.,
A Century of DNA (Cambridge: M.I.T. Press). *Human Nature*, *1*
(June), 18-23.

Turbin, N. V. 1975. "Genetic Engineering: Reality, Perspectives
    and Dangers." *Genetic Engineering and Ecology*, 1-15.
    Arlington, Va.: U.S. Joint Publications Research Service.
    Translated from *Voprosy Filosofii* (Russian), *1*, 47-56.

        Detailed account of the technology and the course of the
        debate in the U.S., intended to "draw the attention of
        Soviet philosophers to consider the social and ethical
        problems of genetic engineering."

Valentine, Raymond C. 1978. "Genetic Blueprints for New Plants."
    The Sciences, *18*, 10-13.

Wald, George. 1976. "The Case Against Genetic Engineering."
    The Sciences, *16* (September/October), 6-11.

    See Ptashne (1976) III for a response.

Watson, James D. 1977. "An Imaginary Monster." *Bulletin of the
    Atomic Scientists*, *33* (May) 12-13.

_____. 1977a. "In Defense of DNA." *The New Republic*, *176*
    (June), 11-14.

Zimmerman, Burke K. 1977. "Self-Discipline or Self-Deception."
    Man and Medicine, *2*, 120-132.

_____. 1978. "Risk-Benefit Analysis--The Cop-Out of Govern-
    mental Regulation." *Trial*, *14*, 43-47.

_____. To appear. "The Right of Free Inquiry--Should the
    Government Impose Limits?" *Proceedings of a Public Forum on
    Recombinant DNA Research, Indiana University, Bloomington,
    November 10-12, 1977*.

_____. To appear. "Recombinant DNA Legislation." *Proceedings
    of a Workshop/Symposium on Frontiers in Genetics Research,
    University of Illinois, Crete, March 3-4, 1978*.

IV. SELECTED NEWS COVERAGE

Chedd, Graham. 1976. "Threat to U.S. Genetic Engineering." *New Scientist, 71, 14-15.*

    Good coverage of the events in Cambridge.

Culliton, Barbara J. 1975. "Kennedy Pushing for More Public Input in Research." *Science, 188* (June), 1187-1189.

_____. 1976. "Recombinant DNA: Cambridge City Council Votes Moratorium." *Science, 193* (July), 300-301.

Fields, Cheryl M. 1976. "Can Scientists Be Trusted on Hazardous Research?" *The Chronicle of Higher Education, 12* (August 2), 4-5.

Gwynne, Peter. 1976. "Politics and Genes." *Newsweek, 87* (January 12), 50-52.

_____. 1977. "Caution Gene Transplant." *Newsweek, 89* (March 21), 57-8.

Leeper, E. M. 1976. "DNA Guidelines Decision at Hand." *Bio-Science, 26* (April), 289-291.

    Coverage of the February 1976 NIH Forum.

Lewin, Roger. 1977. "U.S. Genetic Engineering in a Tangled Web." *New Scientist, 73* (March), 640-641.

    Coverage of the March 1977 Academy Forum. See NAS 1977 I.

McBride, Gail. 1975. "Gene-Grafting Experiments Produce Both High Hopes and Grave Worries," *Journal of the American Medical Association, 232* (April), 337-342.

Marx, Jean L. 1977. "Nitrogen Fixation: Prospects for Genetic Manipulation." *Science, 196* (May), 638-641.

_____. 1977a. "The New P4 Laboratories: Containing Recombinant DNA." *Science, 197* (September), 1350-1352.

O'Sullivan, Dermot A. 1975. "WHO Seeks Role in Genetic Engineering." *Chemical and Engineering News, 53,* 19.

"Pandora's Box of Genes." *The Economist,* March 5, 1977, pp. 82-3.

Parry, Renee-Marie. 1975. "The Promethean Situations: 'The Applications and Limitations of Genetic Engineering—The Ethical Implications,' Davos, 10-12 October 1974." *Futures, 7,* 169-173.

Pollack, Allan. 1977. "Engineering Within the Guidelines."
Trends in Biological Science, 2 (June), N137-8.

  Meeting report from the 9th Miami Winter Symposia (January
  1977). See Scott and Werner 1977 I.

Powledge, Tabitha M. 1977. "Recombinant DNA: The Argument
Shifts." *Hastings Center Report, 7* (April), 18-19.

_____. 1977a. "Recombinant DNA: Backing Off on Legislation:
Why Scientists and Congressmen Changed their Minds."
Hastings Center Report, 7 (December), 8-10.

Russell, Cristine. 1974. "Weighing the Hazards of Genetic
Research: A Pioneering Case Study." *BioScience, 12* (December),
691-694, 744.

_____. 1975. "Biologists Draft Genetic Research Guidelines."
BioScience, 25 (April), 237-240, 277-278.

Seidman, Aaron. 1978. "The U.S. Senate and Recombinant DNA
Research." *Newsletter on Science, Technology and Human
Values, 22* (January), 30-32.

  Report on hearings of the Senate Subcommittee on Science
  Technology and Space, November 2, 8, and 10, 1977, with
  an analysis of the current state of legislation.

_____. 1978. "Legislative Report: The U.S. House of Represent-
atives and DNA." *Newsletter on Science, Technology, and Human
Values, 23,* 23-24.

Steinfels, Peter. 1976. "Biomedical Research and the Public: A
Report From the Airlie House Conference." *The Hastings
Center Report, 6* (June), 21-25.

  See Powledge and Dach (1977) II.

Wade, Nicholas. 1974. "Genetic Manipulations: Temporary Embargo
Proposed on Research." *Science, 185* (July), 332-334.

_____. 1975. "Genetics: Conference Sets Strict Controls to
Replace Moratorium." *Science, 187* (March), 931-935.

_____. 1975a. "Recombinant DNA: NIH Groups Stirs Storm By
Drafting Laxer Rules." *Science, 190* (November), 767-769.

_____. 1975b. "Recombinant DNA: NIH Sets Strict Rules to
Launch New Technology." *Science, 190* (December), 1175-1179.

_____. 1976. "Recombinant DNA: Guidelines Debated at Public
Hearing." *Science, 191* (February), 834-836.

_____. 1976a. "Recombinant DNA: The Last Look Before the Leap." *Science, 192* (April), 236-238.

_____. 1976b. "Recombinant DNA: Chimeras Set Free Under Guard." *Science, 193* (July), 215-217.

_____. 1976c. "Recombinant DNA at White House." *Science, 193* (August), 468.

_____. 1976d. "Recombinant DNA: A Critic Questions the Right to Free Inquiry." *Science, 194* (October), 303-306.

_____. 1976e. "Recombinant DNA: New York State Ponders Action to Control Research." *Science, 194* (November), 705-706.

_____. 1977a. "Gene Splicing: Cambridge Citizens OK Research But Want More Safety." *Science, 195* (January), 268-269.

_____. 1977b. "Dicing With Nature: Three Narrow Escapes." *Science, 195* (January), 378.

_____. 1977c. "Gene-Splicing: Critics of Research Get More Brickbats than Bouquets." *Science, 195* (February), 466-469.

_____. 1977d. "Gene-Splicing: At Grass-Roots Level a Hundred Flowers Bloom." *Science, 195* (February), 558-560.

_____. 1977e. "DNA: Laws, Patents, and a Proselyte." *Science, 195* (February), 762.

_____. 1977f. "Gene-Splicing: Congress Starts Framing Law for Research." *Science, 196* (April), 39-40.

_____. 1977g. "Gene Splicing Preemption Rejected." *Science, 196* (April), 406.

_____. 1977h. "Gene-Splicing: Senate Bill Draws Charges of Lysenkoism." *Science, 197* (July), 348-350.

_____. 1977i. "Recombinant DNA: NIH Rules Broken in Insulin Gene Project." *Science, 197* (September), 1342-1345.

_____. 1977j. "Confusion Breaks Out Over Gene Splice Law." *Science, 198* (October), 176.

_____. 1978. "Gene-Splicing Rules: Another Round of Debate." *Science, 199* (January), 30-33.

_____. 1978a. "Harvard Gene Splicer Told to Halt." *Science, 199* (January), 31.

Weinberg, Janet. 1975. "Asilomar Decision: Unprecedented Guidelines for Gene-Transplant Research." *Science News, 107* (March 8), 148-149.

_____. 1975. "Decision at Asilomar." *Science News, 107* (March 22), 194-196.

Wheeler, Harvey. 1975. "The Regulations of Scientists: Report
from Davos." *The Center Magazine,* 8 (January/February), 73-
77.

Wright, Susan. 1977. "Recombinant DNA Technology: Who Shall
Regulate?" *Bulletin of the Atomic Scientists,* 33 (October),
4-5.

V. POPULAR MAGAZINES AND TRADE JOURNALS-- FEATURE ARTICLES

Baltimore, David. 1976. "Genetic Research . . . Safe or Not?
U.S. Guidelines Give Protection." *Boston Globe,* September 6,
p. 41.

   See Sinsheimer (1976) III.

_____. 1977. "The Gene Engineers." *TV Guide,* March 12.

Bennett, William, and Gurin, Joel. 1976. "Home Rule and the
Gene." *Harvard Magazine,* 79 (October), 14-19+.

   Focuses on the events in Cambridge.

_____. 1977. "Science that Frightens Scientists: The Great
Debate over DNA." *The Atlantic Monthly,* February, pp. 43-62.

   Cover story.

Cavalieri, Liebe F. 1976. "New Strains of Life--Or Death."
New York Times Magazine, August 22, 8-9+. Reprinted in
Kennedy Hearings 1976 II, 9-18.

Crossland, Janice. 1976. "Hands On the Code." *Environment,* 18
(September), 6-16.

Durden-Smith, Jo. 1977. "The Promise of Chi 1977; Roy Curtiss
III's Answer to the DNA Dilemma." *Quest/77*, November/December, pp. 2-10.

"Fruits of Gene-Juggling: Blessing or Curse?" (MWN 1976). *Medical
World News*, *17* (October), 45-46.

Gies, Joseph C. 1977. "Regents Make History: The DNA Decision."
*Association of Governing Boards of Universities and Colleges,
Reports*, *19* (July/August), 3-11.

   Account of debate at the University of Michigan.

Golden, Frederic. 1977. "Tinkering With Life." *Time*, *109*
(April 18), 32-45.

   Cover story.

Hopson, Janet L. 1977. "Recombinant Lab for DNA and My 95 Days
In It." *Smithsonian*, *8* (June), 54-63.

Judson, Horace F. 1975. "Fearful of Science." *Harper's Magazine*, *250*, Part I (March), 32-41; Part II (June), 70-76.

Lubow, Arthur. 1977. "Playing God with DNA." *New Times*,
January 7, pp. 48-63.

McCaull, Julian. 1977. "Research in a Box: Escape of New Life
Forms." *Environment*, *19* (April), 31-37.

Randal, Judith. 1977. "Life From the Labs: Who Will Control the
New Technology?" *The Progressive*, *41* (March), 16-20.

_____. 1977a. "Who Will Oversee the Gene Jugglers?"
Change, *9* (May), 54-5.

Rifkin, Jeremy. 1977. "DNA: Have the Corporations Already
Grabbed Control of New Life Forms?" *Mother Jones*, February/
March, pp. 23-26+.

_____. 1977a. "One Small Step Beyond Mankind." *The Progressive*, *41* (March), 21.

_____, and Howard, Ted. 1977. "Who Should Play God?" *The
Progressive*, *41* (December), 16-22.

Rogers, Michael. 1975. "The Pandora's Box Congress." *Rolling
Stone*, June 19, pp. 36-42.

Sinsheimer, Robert L. 1976. "Genetic Research . . . Safe or Not?
    Extreme Caution Called For." *Boston Globe*, September 6, p. 41.

        See Baltimore (1976) V, and see Sinsheimer (several) III.

VI. TECHNICAL PAPERS

Three distinct lines of research, which were pursued in the
late 1960s, provided the foundation for rDNA research: (i) the
work on restriction as a phenomena whereby incoming fragments of
DNA are degraded by the host, (ii) the work on extra chromosomal
inheritance demonstrating that plasmid DNA replicates independently of the host's DNA, and (iii) the work on DNA ligase, a DNA
repair enzyme.

Today, much of the research on DNA involves rDNA techniques.
Consequently, the literature in this area is accumulating rapidly.
In this section we begin with some of the early papers which were
historically important, because they utilized results from these
various lines of research to produce this molecular technology.
The final two papers report successful cloning of, respectively,
prokaryotic and eukaryotic DNA in *E. coli*.

Kelly, Thomas J., and Smith, Hamilton O. 1970. "A Restriction
    Enzyme from Hemophilus Influenzae II. Base Sequence of the
    Recognition Site." *Journal of Molecular Biology*, *51*, 393-409.

        This demonstrated that a purified restriction endo-
        nuclease will always cleave a certain nucleotide base
        sequence.

Mandel, M., and Higa, A. 1970. "Calcium-dependent Bacteriophage
    DNA Infection." *Journal of Molecular Biology*, *53*, 159-162.

        First demonstration of the transforming properties of
        E. coli, whereby foreign DNA is integrated with host DNA.

Cohen, Stanley N.; Chang, Annie, C.Y.; and Hsu, Leslie. 1972.
    "Nonchromosomal Antibiotic Resistance in Bacteria: Genetic
    Transformations of *E. coli* by R-Factor DNA." *Proceedings of
    the National Academy of Sciences, USA*, *69* (August), 2110-
    2114.

Demonstrates transforming properties of *E. coli* using plasmid DNA.

Mertz, Janet E., and Davis, Ronald W. 1972. "Cleavage of DNA by R$_I$ Restriction Endonuclease Generates Cohesive Ends." *Proceedings of the National Academy of Sciences, USA, 69* (November), 3370–3374.

Herbert Boyer's group had identified and purified a restriction endonuclease, EcoRI. This paper reports that the DNA cleaved by EcoRI possesses "sticky ends," which allow foreign DNA to be inserted into host DNA.

Hedgpeth, Joe; Goodman, Howard M.; and Boyer, Herbert W. 1972. "DNA Nucleotide Sequence Restricted by the RI Endonuclease." *Proceedings of the National Academy of Sciences, USA, 69* (November), 3448–3452.

This paper reports the specific nucleotide sequence cleaved by EcoRI.

Cohen, Stanley N.; Chang, Annie, C. Y.; Boyer, Herbert W.; and Helling, Robert B. 1973. "Construction of Biologically Functional Bacterial Plasmids *In Vitro*." *Proceedings of the National Academy of Sciences, USA, 70* (November), 3240–3244.

EcoRI is used with an *E. coli* host to clone plasmid DNA.

Morrow, John F.; Cohen, Stanley N.; Chang, Annie C. Y.; Boyer, Herbert W.; Goodman, Howard M.; and Helling, Robert B. 1974. "Replication and Transcription of Eukaryotic DNA in *E. coli*." *Proceedings of the National Academy of Sciences USA, 71* (May), 1743–1747.

Xenopus (frog) DNA is cloned in *E. coli*.

VII. SUGGESTIONS FOR FURTHER READING

This last section is intended to provide a starting point for the reader interested in pursuing, beyond his or her own discipline, some of the issues underlying the rDNA controversy. Here are a few works which represent some of the different perspectives from which this complex of problems has been approached. None directly treat rDNA.

A. Social Perspectives

We mention first, some works which treat the general relationship of science, and the scientist, to our society.

Edsall, John T.   1975.   "Scientific Freedom and Responsibility."
    Science, 188 (May), 687–693.
        Abbreviated version of the report of the AAAS Committee
        on Scientific Freedom and Responsibility.

Holton, Gerald, and Blanpied, William A., eds. 1976. *Science and Its
    Public: The Changing Relationship*. Vol. XXXIII of *Boston
    Studies in the Philosophy of Science*. Edited by R. S. Cohen
    and M. Wartofsky. Dordrecht, Holland: D. Reidel.

Ravetz, Jerome R.   1971.   *Scientific Knowledge and Its Social
    Problems*. New York: Oxford University Press.
        An essay in the sociology of knowledge as it applies
        to science, and the scientific community.

Two collections of papers focus on the interrelationship of
contemporary biology and society:

Fuller, Watson, ed.   1971.   *The Social Impact of Modern Biology*.
    London: Routledge and Kegan Paul.

Handler, Philip, ed.   1970.   *Biology and the Future of Man*.
    London: Oxford University Press.

The growing field of Bioethics takes on issues such as,
abortion, experimentation with human subjects, euthanasia, human
genetic counseling and engineering, and the nature of health and
disease. A journal, *The Hastings Center Report*, is devoted
entirely to these subjects. There are several good collections of
readings available. We mention two:

Almeder, Robert S., and Humber, James , eds.   1976.   *Biomedical
    Ethics and the Law*. New York: Plenum Press.

Beauchamp, Tom L. and Walters, LeRoy, eds.   1978. *Contemporary Issues in Bioethics*. California: Dickenson.

B. *Legal Perspectives*

The major reference work on the relationship of the law to scientific and technological developments is:

Tribe, Laurence H. 1973. *Channeling Technology Through the Law*. Chicago: Bracton Press.

We mention several articles which address the viability of risk-benefit analysis and the general problem of utilizing scientific information in legal and social decisions.

Gelpe, Marcia R., and Torlock, A. Dan. 1974. "The Uses of Scientific Information in Environmental Decision-making." *Southern California Law Review, 48, 371-427*.

Green, Harold P. 1973. "Genetic Technology: Law and Policy for the Brave New World." *Indiana Law Journal, 48, 559-576*.

_____. 1975. "The Risk-Benefit Calculus in Safety Determinations." *George Washington Law Review, 43, 791-807*.

>   See also Handler (1975) VII B.

Handler, Phillip. 1975. "A Rebuttal: The Need for a Sufficient Scientific Base for Governmental Regulations." *George Washington Law Review, 43, 808-813*.

>   See Green (1975) VII B.

Lederberg, Joshua. 1972. "The Freedoms and the Control of Science: Notes from the Ivory Tower." *Southern California Law Review, 45, 596-614*.

>   It would be as wrong as it is futile for scientists to resist contemporary inquiry about the social merits of scientific and technological progress, or to ignore concrete proposals. This generally explores the relationship of contemporary science to society, and encourages further exploration.

C. Philosophical Perspectives

There are several general portrayals of contemporary man's position in society, and the relationship among technological and scientific innovation, society, and the role of the individual. Two major works, by contributors to this volume, are:

Callahan, Daniel. 1975. *The Tyranny of Survival*. New York: Macmillan.

Jonas, Hans. 1974. *Philosophical Essays: From Ancient Creed to Technological Man*. Englewood Cliffs: Prentice Hall.
> See, in particular, part one: "Science, Technology and Ethics."

A brief essay defending a humanistic perspective in science, by a philosopher of science who has authored a book on rDNA (Goodfield (1977) I) is:

Goodfield, June. 1977a. "Humanity in Science: A Perspective and a Plea." *Science, 198* (November), 580-585.

Philosophy of Biology applies philosophical methods to problems in the foundations of biology as a science. The concern is with topics such as, holism and reductionism, teleology, biological explanation, and taxonomy.

Beckner, Morton. 1959. *The Biological Way of Thought*. Berkeley: University of California, 1968.

Hull, David. 1974. *Philosophy of Biological Science*. Englewood Cliffs: Prentice Hall.

Medawar, P. B., and Medawar, J. S. 1977. *The Life Sciences: Current Ideas of Biology*. New York: Harper and Row.

Ruse, Michael. 1973. *The Philosophy of Biology*. London: Hutchinson University Library.

Two good collections in this area are:

Ayala, Francisco J., and Dobzhansky, Theodosius, eds.   1974.
    *Studies in the Philosophy of Biology: Reduction and Related
    Problems*. Berkeley: University of California Press.

Grene, Marjorie, and Mendelsohn, Everett, eds.   1976. *Topics in
    the Philosophy of Biology*. Vol. XXVII of *Boston Studies in the
    Philosophy of Science*. Edited by R. S. Cohen and M. Wartofsky.
    Dordrecht, Holland: D. Reidel.

D. *Biologists' Perspectives*

Important insights into the state of a science, and into the
role of the scientist in science and society, can be gained
through the scientist's own writings. We list some classic
statements of biologists reflecting on their own discipline.

Dawkins, Richard. 1976. *The Selfish Gene*.   Oxford: Oxford Uni-
    versity Press.

Dobzhansky, Theodosius. 1967. *The Biology of Ultimate Concern*.
    New York: New American Library.

Hardin, Garrett. 1959. *Nature and Man's Fate*.   New York: New
    American Library.

Huxley, Aldous. 1932. *Brave New World*.   New York: Harper and
    Row.

Monod, Jacques. 1971. *Chance and Necessity*.   New York: Alfred A.
    Knopf.

Thomas, Lewis. 1974. *The Lives of a Cell*.   Toronto: Bantam Books.

Waddington, C. H. 1960. *The Ethical Animal*.   London: George
    Allen and Unwin.

Watson, James D. 1968. *The Double Helix: A Personal Account of
    the Discovery of the Structure of DNA*.   New York: New American
    Library.

E. Bibliographical Sources

Walters, LeRoy, et al. 1975-. *Bibliography of Bioethics*, pub-
    lished annually. Michigan: Gale Research Company.

    Excellent comprehensive annotated bibliography and general
    guide to the literature in bioethics.

We also refer the reader to the *Project on the Development of
Recombinant DNA Research Guidelines*, under the direction of Charles
Wiener of the M.I.T. Oral History Program.

Index

365